T0230449

Emerging Technologies in Manufacturing

Matthew N. O. Sadiku
Abayomi J. Ajayi-Majebi
Philip O. Adebo

Emerging Technologies in Manufacturing

 Springer

Matthew N. O. Sadiku
Prairie View A&M University
Prairie View, TX, USA

Abayomi J. Ajayi-Majebi
Central State University
Wilberforce, OH, USA

Philip O. Adebo
Texas Southern University
Houston, TX, USA

ISBN 978-3-031-23158-2   ISBN 978-3-031-23156-8   (eBook)
https://doi.org/10.1007/978-3-031-23156-8

This Springer imprint is published by the registered company Springer Nature Switzerland AG
The registered company address is: Gewerbestrasse 11, 6330 Cham, Switzerland

Dedicated to:

Janet Sadiku

Chief Alfred Bamidele, Chief (Mrs.) Victoria Adejoke Ajayi-Majebi, and Mr. Olusegun Ajayi-Majebi (All Deceased)

Victoria Adebo

Preface

One of the most vital industries worldwide is the manufacturing industry. The manufacturing industry is a cornerstone of national economy and people's livelihood. Manufacturing is regarded as a prime generator of wealth and is critical in establishing a sound basis for economic growth. From craft production to cloud manufacturing, manufacturing has become an ever-increasingly complex process, relying on many new technologies.

Manufacturing is the way of transforming resources into products or goods which are required to cater to the needs of the society. It refers to the entire product life cycle: product design, production planning, production, distribution, field service, and reclamation. It consists of a set of processes, machines, and factories where raw materials are transformed into products. There are different types of manufacturing including biopharmaceutical manufacturing, chemical manufacturing, discrete manufacturing, advanced manufacturing, smart manufacturing, cloud manufacturing, offshore manufacturing, lean manufacturing, green manufacturing, distributed manufacturing, predictive manufacturing, and computer-aided manufacturing. The leading manufacturing industries in the United States include steel, automobiles, chemicals, food processing, consumer goods, aerospace, and mining.

Traditional manufacturing companies currently face several challenges such as rapid technological changes, inventory problem, shortened innovation, short product life cycles, volatile demand, low prices, highly customized products, and ability to compete in the global markets. Modern manufacturing is highly competitive due to globalization and fast changes in the global market. Globalization has had a huge impact on economic, social, and political changes. The manufacturing industry has become more competitive due to globalization. To survive from the global market, manufacturing enterprises should reduce the product cost and increase the productivity. Manufacturing systems are evolving into a new generation of data-driven systems that must adapt to these changes in a timely fashion. In today's manufacturing, technology is the determining factor for securing competitive advantage. The industry has always been receptive to adopting new technologies such as drones, industrial robots, artificial intelligence (AI), virtual reality, and Internet of Things.

These technologies have come to rescue many manufacturing organizations drowned in the deluge of data, data explosion, and information overload. Manufacturers and policy-makers must keep track of more than 60 technologies and philosophies shaping manufacturing today. For each technology, they must assess its readiness, adoption level, and its most relevant applications in manufacturing and the key barriers to further adoption.

This book reviews emerging technologies in manufacturing. These technologies include artificial intelligence, smart manufacturing, lean manufacturing, robotics, automation, 3D printing, nanotechnology, industrial Internet of Things, digital twins or cyber-physical systems (CPS) and augmented reality. The use of these technologies will have a profound impact on the manufacturing industry. They have the potential to transform manufacturing as we know it. They will decrease labor costs, improve efficiency, and reduce waste, making factories to be more environment-friendly. They should be at the core of any manufacturing upgrading effort. To remain competitive, forward-thinking manufacturers should embrace these emerging manufacturing technologies.

This book is designed to provide an introductory treatment on newly emerging technologies for manufacturing. It consists of 19 chapters. Each chapter addresses a single emerging technology in depth and describes how manufacturing organizations are adopting the technology. Chapter 1 provides an introduction on emerging technologies in manufacturing. Chapter 2 covers artificial intelligence (AI), which is the cognitive science that deals with intelligent machines which are able to perform tasks heretofore only performed by human beings. Today, the manufacturing industry is using artificial intelligence and machine learning to streamline every phase of production and increase productivity. In Chap. 3, we discuss the uses of robotic automation in manufacturing. Automation can assist the manufacturing industry by reducing labor costs substantially, increase quantity or generate more output for a given period of time, improve product quality, and mitigate risk. Chapter 4 covers smart manufacturing, which has gained significant impetus as a breakthrough technological development that can transform the landscape of manufacturing today and tomorrow. It aims to take advantage of advanced information and manufacturing technologies to enable flexibility in physical processes to address a dynamic and global market.

Chapter 5 presents Industry 4.0 as a manufacturing initiative. Industry 4.0 enables the digitalization of manufacturing sector with built-in sensing devices virtually in all manufacturing components, products, and equipment. It has been involved in smart transportation and logistics, smart buildings, oil and gas, smart healthcare, smart cities, to name a few. Chapter 6 introduces Industrial Internet of Things (IIoT), which refers to particular applications of IoT to industrial environments. IIoT is a new industrial ecosystem that combines intelligent and autonomous machines in support of digital twinning and cyberphysical systems. It is bringing about a world where smart, connected embedded systems and products operate as part of larger systems. Chapter 7 covers big data, which refers to the large volume of structured and unstructured data with the potential to be mined for information. The manufacturing industry has joined the big data bandwagon as well because data have long been the lifeblood of manufacturing. Big data analytics can provide the manufacturing industry the need to succeed in an increasingly complex environment. In Chap. 8, additive manufacturing is presented. It adds material to create complex and intricate objects in the context of less material waste, while simultaneously cutting industrial resources waste. While traditional production systems proceed by subtraction, additive manufacturing involves fabricating a 3D object by successively adding material. Compared to subtractive manufacturing, additive manufacturing can compete with or beat traditional manufacturing processes in terms of cost and adaptability.

In Chap. 9, we address green manufacturing, which is a new trend for the future development of the manufacturing industry. Green manufacturing refers to modern manufacturing that makes products without pollution. It addresses a wide range of environmental and sustainability issues including resource selection, transportation, manufacturing process, and pollution. Chapter 10 dwells on sustainable manufacturing, which is manufacturing products through economically sound processes that minimize negative environmental impacts while conserving energy and natural resources. The goal of sustainable manufacturing is to minimize waste, maximize resource efficiency, and reduce the environmental impact of manufacturing. Chapter 11 presents lean manufacturing, which is all about doing more with less by applying "lean thinking." It is a systematic method that focuses on eliminating or reducing waste within manufacturing systems without sacrificing productivity. Chapter 12 introduces the readers to distributed manufacturing, which refers to decentralized fabrication of parts in smaller factories or homes that are local to end-users. This manufacturing-as-a-service model is achieved by making small orders and prototyping affordable and building their production facilities near the places of consumption.

Chapter 13 deals with cloud manufacturing, which is also known as manufacturing-as-a-Service. It is a newly emerging approach of adopting well-known basic concepts from cloud computing to manufacturing processes and delivering shared, ubiquitous, on-demand manufacturing services. It refers to a customer-centric approach to enabling ubiquitous, convenient, on-demand network access to a shared pool of manufacturing resources and capabilities. In Chap. 14, we cover nanomanufacturing, which is the manufacturing of objects or material with dimensions between 1 and 100 nanometers. It focuses on developing scalable, high-yield processes for the production of materials, structures, devices, and systems at the nanoscale. Chapter 15 introduces the concept of ubiquitous manufacturing. Ubiquitous manufacturing is an application of ubiquitous computing which is a concept where computing is made to appear everywhere and anywhere. It emphasizes the mobility and dispersion of manufacturing resources, products, and users. Chapter 16 addresses offshoring manufacturing, which takes place when production operations are performed in another country, from one country to a low-cost country. It is considered an option when a company wants to reduce operating expenses while maintaining high-quality product.

In Chap. 17, we discuss cyber manufacturing, a modern manufacturing approach that utilizes cyber-physical systems also popularized as digital twins. It may be regarded as the convergence of cyber-physical systems, systems engineering, and manufacturing innovation. It is a new transformative concept that involves the translation of data from interconnected systems into predictive and prescriptive operations, leveraging the advantages of augmented reality. Chapter 18 dwells on biomanufacturing, the production of biological products from living cells. It is an emerging manufacturing process that utilizes biological systems to produce commercially important biomaterials and biomolecules. It is used in medicine, food industry, and industrial applications. Chapter 19 addresses the future of manufacturing, which is undeniably digital and will be shaped by a combination of advancements in technology. Other topics on emerging technologies in manufacturing, not covered as chapters in the book or not covered at all, are listed in Appendix A.

The book fills an important niche for manufacturing. It is a comprehensive, introductory text on the issues, ideas, theories, and problems of emerging technologies in manufacturing. It provides an introduction to these technologies so that beginners can understand the technologies, their increasing importance, and their developments in contemporary time. It is a must-read book for anyone who wants to be updated about emerging technologies.

Prairie View, TX, USA   Matthew N. O. Sadiku
Wilberforce, OH, USA   Abayomi J. Ajayi-Majebi
Houston, TX, USA   Philip O. Adebo

Contents

1 Introduction 1
  1.1 Introduction 1
  1.2 Some Emerging Technologies 2
  1.3 Benefits 8
  1.4 Challenges 9
  1.5 Global Aspect of Emerging Technologies in Manufacturing. 9
  1.6 Conclusion 10
  References. 10

2 Artificial Intelligence in Manufacturing 13
  2.1 Introduction 13
  2.2 Overview on Artificial Intelligence 14
  2.3 AI in Manufacturing 17
  2.4 Applications of AI in Manufacturing. 19
  2.5 Benefits 24
  2.6 Challenges 25
  2.7 Global AI in Manufacturing. 26
  2.8 Future of AI in Manufacturing. 27
  2.9 Conclusion 29
  References. 30

3 Robotic Automation in Manufacturing 33
  3.1 Introduction 33
  3.2 What Is a Robot? 34
  3.3 Overview on Automation. 35
  3.4 Robotics in Manufacturing 37
  3.5 Manufacturing Applications. 39
  3.6 Benefits 43
  3.7 Challenges 44
  3.8 Global Adoption of Robotic Automation. 45
  3.9 Conclusion 45
  References. 46

4   Smart Manufacturing 49
   4.1   Introduction 49
   4.2   Overview of Traditional Manufacturing 50
   4.3   Concept of Smart Manufacturing. 51
   4.4   Characteristics of Smart Manufacturing 52
   4.5   Goals of Smart Manufacturing. 54
   4.6   Enabling Technologies. 54
   4.7   Applications. 56
   4.8   Smart Factory. 58
   4.9   Benefits 60
   4.10   Challenges 62
   4.11   Global Adoption of Smart Manufacturing 62
   4.12   Conclusion 64
   References. 65

5   Industry 4.0 67
   5.1   Introduction 67
   5.2   Industrial Revolutions 68
   5.3   Overview on Industry 4.0. 69
   5.4   Principles and Features 72
   5.5   Applications. 74
   5.6   Benefits 76
   5.7   Challenges 77
   5.8   Global Adoption of Industry 4.0 79
   5.9   Conclusion 80
   References. 80

6   Industrial IOT in Manufacturing 83
   6.1   Introduction 83
   6.2   Overview of Internet of Things 84
   6.3   Industrial Internet of Things 85
   6.4   Applications. 87
   6.5   Benefits 89
   6.6   Challenges 90
   6.7   Global Adoption of IIOT 92
   6.8   Conclusion 93
   References. 93

7   Big Data in Manufacturing 95
   7.1   Introduction 95
   7.2   Big Data Characteristics 96
   7.3   Big Data Analytics. 98
   7.4   Big Data in Manufacturing 99
   7.5   Applications. 100
   7.6   Benefits 102
   7.7   Challenges 103

7.8   Global Adoption of Big Data in Manufacturing   105
7.9   Conclusion   106
References   106

8   Additive Manufacturing   109
8.1   Introduction   109
8.2   What Is Additive Manufacturing?   110
8.3   Additive Manufacturing Processes   112
8.4   Applications   113
8.5   Benefits   115
8.6   Challenges   116
8.7   Global Adoption of Additive Manufacturing   118
8.8   Conclusion   119
References   120

9   Green Manufacturing   123
9.1   Introduction   123
9.2   Traditional Manufacturing   124
9.3   What Is Green Manufacturing?   126
9.4   Motivations for Green Manufacturing   127
9.5   Characteristics of GM   128
9.6   Green Manufacturing Practices   129
9.7   Green Manufacturing Strategies   131
9.8   Applications of Green Manufacturing   131
9.9   Awareness of GM   133
9.10   Benefits   133
9.11   Challenges   135
9.12   Global Impact of Green Manufacturing   136
9.13   Conclusion   138
References   139

10   Sustainable Manufacturing   141
10.1   Introduction   141
10.2   The Concept of Sustainability   142
10.3   What Is Sustainable Manufacturing?   144
10.4   Sustainable Approaches   146
10.5   Sustainable Manufacturing Examples   147
10.6   Benefits   149
10.7   Challenges   150
10.8   Global Adoption of Sustainable Manufacturing   151
10.9   Conclusion   153
References   154

11   Lean Manufacturing   157
11.1   Introduction   157
11.2   Types of Waste   158
11.3   Concept of Lean Manufacturing   159

11.4   Lean Manufacturing Principles 160
11.5   Relationship of Lean and Other Concepts 162
11.6   Why Are Companies Using Lean Manufacturing? 163
11.7   Applications 164
11.8   Benefits 166
11.9   Challenges 168
11.10 Global Adoption of Lean Manufacturing 169
11.11 Conclusion 171
References 172

12   **Distributed Manufacturing** 175
12.1   Introduction 175
12.2   Concept of Distributed Manufacturing 176
12.3   Distributed Manufacturing Systems 176
12.4   Features of Distributed Manufacturing 177
12.5   Applications of Distributed Manufacturing 178
12.6   Benefits 179
12.7   Challenges 180
12.8   Global Adoption of Distributed Manufacturing 180
12.9   Conclusion 183
References 183

13   **Cloud Manufacturing** 185
13.1   Introduction 185
13.2   Overview of Cloud Computing 186
13.3   Cloud Manufacturing 188
13.4   Migrating to the Cloud 190
13.5   Applications 191
13.6   Benefits 192
13.7   Challenges 194
13.8   Global Cloud Manufacturing 194
13.9   Future of Cloud Manufacturing 196
13.10 Conclusion 197
References 197

14   **Nanomanufacturing** 201
14.1   Introduction 201
14.2   Overview of Nanotechnology 202
14.3   Nanomanufacturing 203
14.4   Applications 205
14.5   Benefits 207
14.6   Challenges 208
14.7   Global Nanomanufacturing 209
14.8   Future of Nanomanufacturing 211
14.9   Conclusion 212
References 212

15   Ubiquitous Manufacturing.   215
    15.1   Introduction   215
    15.2   Overview on Ubiquitous Computing   216
    15.3   Ubiquitous Manufacturing   218
    15.4   Related Technologies   219
    15.5   UM Characteristics   220
    15.6   Enabling Technologies.   221
    15.7   Applications   222
    15.8   Benefits   224
    15.9   Challenges   226
    15.10 Global Ubiquitous Manufacturing   228
    15.11 Future of Ubiquitous Manufacturing   229
    15.12 Conclusion   230
    References.   230

16   Offshore Manufacturing.   233
    16.1   Introduction   233
    16.2   What Is Offshore Manufacturing?   234
    16.3   Local or Offshore Manufacturing   237
    16.4   Offshoring and Outsourcing.   239
    16.5   Offshore Locations.   240
    16.6   Benefits of Offshoring   242
    16.7   Challenges of Offshoring.   242
    16.8   Conclusion   244
    References.   245

17   Cyber Manufacturing.   247
    17.1   Introduction   247
    17.2   What Is Cyber Manufacturing   248
    17.3   Enabling Technologies.   249
    17.4   Attack on Cyber Manufacturing   251
    17.5   Handling Cybersecurity Risks   253
    17.6   Benefits   254
    17.7   Challenges   254
    17.8   Global Cyber Manufacturing.   255
    17.9   Conclusion   256
    References.   257

18   Biomanufacturing.   259
    18.1   Introduction   259
    18.2   What Is Biomanufacturing?   260
    18.3   Applications   262
    18.4   Benefits   266
    18.5   Challenges   266
    18.6   Global Biomanufacturing.   268
    18.7   Future of Biomanufacturing.   270

        18.8   Conclusion 271
        References 272

19   Future of Manufacturing 275
        19.1   Introduction 275
        19.2   Emerging Manufacturing Technologies 276
        19.3   Future of Technology in Manufacturing 277
        19.4   Factory of the Future 280
        19.5   Challenges 282
        19.6   Future of Global Manufacturing 284
        19.7   Conclusion 285
        References 286

Index 287

About the Authors

Matthew N. O. Sadiku received his B.Sc. degree in 1978 from Ahmadu Bello University, Zaria, Nigeria and his M.Sc. and Ph.D. degrees from Tennessee Technological University, Cookeville, TN, in 1982 and 1984, respectively. From 1984 to 1988, he was an assistant professor at Florida Atlantic University, Boca Raton, FL, where he did graduate work in computer science. From 1988 to 2000, he was at Temple University, Philadelphia, PA, where he became a full professor. From 2000 to 2002, he was with Lucent/Avaya, Holmdel, NJ as a system engineer and with Boeing Satellite Systems, Los Angeles, CA as a senior scientist. He is presently a professor emeritus of electrical and computer engineering at Prairie View A&M University, Prairie View, TX.

He is the author of over 1050 professional papers and over 100 books including standard texts such as *Elements of Electromagnetics* (Oxford University Press, seventh ed., 2018), *Fundamentals of Electric Circuits* (McGraw-Hill, seventh ed., 2021, with C. Alexander), *Computational Electromagnetics with MATLAB* (CRC Press, fourth ed., 2019), *Principles of Modern Communication Systems* (Cambridge University Press, 2017, with S. O. Agbo), and *Signals & Systems: A Primer with MATLAB* (CRC Press, 2016, with W. H. Ali). In addition to the engineering books, he has written Christian books including *Secrets of Successful Marriages*, *How to Discover God's Will for Your Life*, and commentaries on all the books of the New Testament Bible. Some of his books have been translated into French, Korean, Chinese (and Chinese Long Form in Taiwan), Italian, Portuguese, and Spanish.

He was the recipient of the 2000 McGraw-Hill/Jacob Millman Award for outstanding contributions in the field of electrical engineering. He was also the recipient of Regents Professor award for 2012-2013 by the Texas A&M University System. He is a registered professional engineer (PE) and a fellow of the Institute of Electrical and Electronics Engineers (IEEE) "for contributions to computational electromagnetics and engineering education." He was the IEEE Region 2 Student Activities Committee Chairman. He was an associate editor for *IEEE Transactions on Education*. He is also a member of Association for Computing Machinery (ACM) and American Society of Engineering Education (ASEE). His current research interests are in the areas of computational electromagnetic, computer networks, and engineering education. His works can be found in his autobiography, *My Life and Work* (Trafford Publishing, 2017), or his website: www.matthew-sadiku.com. He currently resides in West Palm Beach, Florida. He can be reached via email at sadiku@ieee.org

Abayomi J. Ajayi-Majebi is a licensed professional engineer (PE) registered with the State of Ohio and a full-time professor and past Chairman of the Manufacturing Engineering Department at Central State University (CSU) in Wilberforce, Ohio, where he also teaches Manufacturing Engineering and Nuclear Engineering (Minor) to undergraduate students. He has an earned doctorate in Civil Engineering from The Ohio State University in Columbus, Ohio (1983) in addition to two Master's degrees (1980, 1985). He is a permanent member of American Society of Engineering Education (ASEE) and a senior member/member of seven other professional organizations. He was the 2022 ASEE Co-Program Chair for the Engineering Design Graphics Division (EDGD) and he promotes the interests of ASEE and other engineering professional societies. He had been an adjunct professor at two other Ohio universities, Cedarville University and Wilberforce University. He has supported the education of hundreds of US graduate manufacturing engineers. In support of the professional formation of undergraduate engineering students, he teaches engineering law and engineering ethics to undergraduate engineering students in support of ABET-accreditation goals and objectives to enable them safely navigate the legal and regulatory environments of their future engineering practices. He has performed research in support of over 30 U.S. organizations and has been the author and co-author of numerous publications. He was inducted into the International Educators' Hall of Fame on December 30, 2022, in Anaheim, California.

Dr. Ajayi-Majebi is a past recipient of the Central State University (CSU) Distinguished Faculty Excellence Award for Research (1994) and for Community Service (1995). He is the past recipient of the Outstanding Faculty of the Year Award in the College of Business and Industry (2005). He received the CSU Faculty of the Year Award in the College of Science and Engineering (2012). He is a recent recipient of the Outstanding Presidential Faculty of the Year Award for Service (2021), He is a life-time member of The Ohio State University Alumni Association. He is an Ordained Minister and a member of numerous Christian Organizations across the World and subscribes to the higher Christian values.

As a senior member of the Society of Manufacturing Engineers, he has served as faculty adviser for both the CSU Student Chapter of the Society of Manufacturing Engineers (SME) and the CSU Student Chapter of the National Society for Black Engineers (NSBE). He is an endowed member of the American Nuclear Society and a senior member of the American Society for Quality.

Philip O. Adebo is a faculty member at Texas Southern University. He received his Ph.D. in Electrical and Computer Engineering at Prairie View A&M University, Prairie View, Texas. He is an author and co-author of several papers. His research interests include power systems, renewable energy, microgrids, smart-grid systems, power grid integration, and optimization.

Chapter 1
Introduction

Manufacturing is more than just putting parts together. It's coming up with ideas, testing principles and perfecting the engineering, as well as final assembly.

–James Dyson.

1.1 Introduction

Manufacturing is the way of transforming resources into products or goods which are required to cater to the needs of the society. It constitutes the foundation of any nation's economic development. The manufacturing industry includes many sectors such aerospace industry, food and beverage industry, mining industry, apparel industry, chemical industry, automotive industry, metal and machinery industry, packaging industry, and electronics industry. Manufacturing has been noted as one of the largest industries globally. It is a key component of economic output of any nation because of its contribution to GDP and employment generation. Manufacturing basically impacts economic structures at global, regional, national, and local levels.

Manufacturers are well known for being slow in adopting a new technology. To remain competitive and reach its full potential, the manufacturing organization will need to continue to embrace new technologies. To stay ahead of competitors and win market share in today's quickly morphing environment, manufacturers should embrace change.

What manufacturing looks like has changed drastically within a short time. Manufacturing has evolved from a mechanical process to a set of processes based on the integration of information and technology. Technologies touch on every step of the end-to-end manufacturing process and global value chains. A set of new technologies that promise to boost productivity is emerging. In this chapter, we will address eight sweeping manufacturing technology trends which can disrupt and change how manufacturing companies execute current processes. These emerging technologies will impact the prospects for manufacturing in the early twenty-first century [1].

R&D scientists and engineers are playing a critical role in delivering the latest and greatest technology. They have long realized the benefits and the risks of emerging technologies. Manufacturing organizations will continue to pursue new

M. N. O. Sadiku et al., *Emerging Technologies in Manufacturing*,
https://doi.org/10.1007/978-3-031-23156-8_1

technologies that are better than the status quo. They will invest in technologies that digitize human efforts down to each movement. Their willingness to explore these technologies indicates that digitization in manufacturing is long overdue.

This chapter provides an introduction to emerging technologies in manufacturing. It presents some emerging manufacturing technologies. It addresses the benefits and challenges of the emerging technologies. It considers the global aspect of emerging technologies in manufacturing. The last section concludes with comments.

1.2 Some Emerging Technologies

Here, we consider some emerging technologies that are changing the face of manufacturing. They can disrupt and change how manufacturing companies execute current processes. They will make a lasting impact. We will take a deeper look at the following key technologies:

1. *Artificial Intelligence:* Artificial intelligence (AI) refers to intelligence exhibited by machines or computer systems in being able to perform some tasks that are normally considered to require human intelligence. AI has several functions such as learning, understanding, reasoning, interaction, and decision-making. Technologies that emanate from AI (known as cognitive technologies) include expert systems, fuzzy logic, machine learning, deep learning, computer vision, natural language processing, speech recognition, computer vision, and robotics.

   In particular, machine learning, a branch of AI, uses algorithms to get computers to perform a task effectively without being programmed explicitly. AI and machine learning are widely used in manufacturing for demand forecasting and predictive maintenance. AI is used at various points of the manufacturing process. An area of manufacturing where AI has been applied with great success is aircraft assembly. Manufacturing efficiency and maintaining product quality remain top priorities in some industries such as the information communication and integrated circuit (IC^2) industry, where process and equipment reliability directly influence cost, throughput, and yield [2]. Machine learning can help manufacturers analyze data, automatically discover hidden opportunities, accelerate tedious processes, identify which data insights matter, and reduce operational costs. AI is used at various points in the manufacturing process at BMW's plants to save costs and streamline production.

2. *Robotic Automation:* Automation is replacing dangerous and once-odd and undesirable human jobs. It has helped manufacturing industry remove the need for human intervention. Manufacturers can automate one process at a time. Since robots are designed for different tasks, there are many types of robots including social robots, collaborative robots, industrial robots, gantry robots, palletizing robots, mobile robots, etc. There is an increase in the number of small- and medium-sized manufacturing companies employing collaborative robots (or cobots) because they are faster to deploy and generate returns quicker. Cobots

are programmable through assisted movement and are specifically designed to work side-by-side with humans. Industrial robotics is mainly responsible for eliminating manufacturing jobs. Robotics has become invaluable for monotonous jobs like sorting, scanning, and packaging. Robots are increasingly being adopted in healthcare, retail, food industry, and manufacturing to increase product quality, yield rates, reduce operating costs, improve product reliability, reduce production delays, increase production efficiency, and improve time-to-customer production performance. Manufacturers of all types—industrial biotechnology, drugs, cars, electronics, or other material goods—are switching to robotic automation to remain competitive. The automotive industry is where robotic automation is very useful. The European aircraft manufacturer Airbus uses mobile robots. Robotic automation will increasingly rely on AI. A task that would require 500 workers can be done by just a few robots with only five technicians to service the robots. Manufacturers across sectors (automotive, electronics, biotech, drugs, etc.) rely on robotic automation to remain competitive. A typical setting where robots do all the physical work is illustrated in Fig. 1.1 [3]. NASA intends to send its first mobile robot to the Moon in 2023. With rapid advances in robotics technology, robots are becoming cheaper, smarter, and more efficient.

3. *Additive Manufacturing/3D Printing:* Without doubt, 3D printing will change the face of manufacturing forever. 3D printing describes the process of manufacturing parts or products layer by layer, as opposed to subtractive or material removal manufacturing. Compared to conventional manufacturing processes, 3D printers can create complex geometries, offering greater freedom for design-

Fig. 1.1 A typical setting where robots do all the physical work [3]

ers. It can also minimize material waste and reduce the time-to-market. Some designers unknowingly still perform their design using conventional processes and avoid technologies like 3D printing in manufacturing. Various materials such as rubber, nylon, plastic, glass, metal, synthetic gems, cellophane, Bakelite, stainless steels, synthetic rubber, Teflon, silicon transistors, float glass, liquid crystals, optical fibers, etc. can be used to print objects. Although it is now possible to produce almost any component using metal, plastic, and human tissue, there are currently no super printer that can do all things, like printing metal and also plastic parts. Manufacturers will benefit from faster, less-expensive production due to 3D printing. 3D printing also makes it possible for tooling to be cost-effectively completed on-site, in a short amount of time. The increasing applications of 3D printing in manufacturing are giving rise to manufacturing as a service (MaaS). The 3D printing industry is growing at astronomical rates [4]. Figure 1.2 depicts 3D printing in manufacturing [5].

4. *Industrial Internet of Things*: The Internet of Things (IoT) is a revolutionary manufacturing technology. It allows physical objects (devices, vehicles, buildings, equipment, etc.) to collect and exchange data. It allows electronic devices to connect to each other within an existing Internet infrastructure. The interconnection of devices is to achieve a variety of goals including communication between the devices, cost reduction, increased efficiency, and improved safety. IoT's existence is primarily due to three factors: widely available Internet access, smaller sensors, and cloud computing. IoT is a part of the predictive technology that manufacturers are using with great success. For manufacturers, IoT means having more data available for monitoring and improving operations. Manufacturers have applied IoT to increase operational efficiency, while some

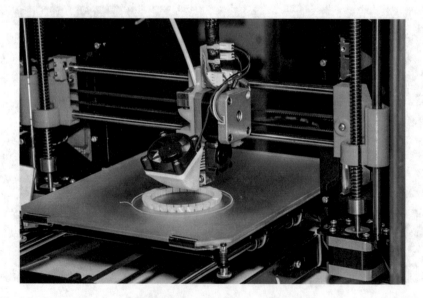

Fig. 1.2 3D printing in manufacturing [5]

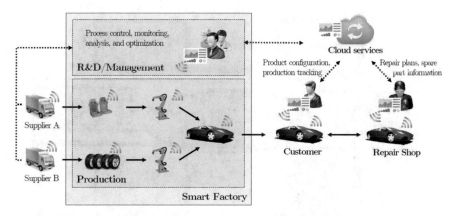

Fig. 1.3 A typical industrial Internet of Things [7]

manufacturers have not implemented IoT due to cybersecurity concerns. The industrial Internet of Things (IIoT) is the extension of IoT in industrial settings. It refers to the application of the Internet of Things (IoT) across several industries such as manufacturing, logistics, oil and gas, transportation, energy/utilities, chemical, aviation, and other industrial sectors [6]. It is an amalgamation of various technologies such as IoT, machine learning, big data, cloud integration, and machine automation. Its aim is to transform the production modes and improve the production efficiency. A typical industrial Internet of Things is shown in Fig. 1.3 [7]. Manufacturers who have not implemented IIoT should realize that they are already behind their competitors. Edge computing, or computing done closer to the sensor, is a new trend within IIoT platform. Edge computing offers several benefits to manufacturing, such as increased efficiency, lower costs, and efficient bandwidth. Edge computing clears the path for the autonomous factory.

5. *Smart Manufacturing:* Smart manufacturing (SM) refers to the integration and networking of manufacturing system elements within a factory, enterprise, and supply chain. Other initiatives such as Industry 4.0, cyber-physical production systems, smart factory, intelligent manufacturing, and advanced manufacturing are frequently used synonymously with smart manufacturing. SM is characterized by the integration of new-generation information communication technology (ICT) and the manufacturing industry. Manufacturers can use smart manufacturing to improve productivity and efficiency. The movement toward smart-manufacturing facilities and digitalized value chains is having a major impact on manufacturing. Smart manufacturing involves new technologies such as industrial Internet of Things (IIoT), artificial intelligence (AI), and robotic automation. Smart manufacturing, which is the fourth revolution in the manufacturing arena, has its own identity captured in the following six pillars [8], shown in Fig. 1.4: (1) manufacturing technology and processes, (2) materials, (3) data, (4) predictive engineering, (5) sustainability, (6) resource sharing and networking.

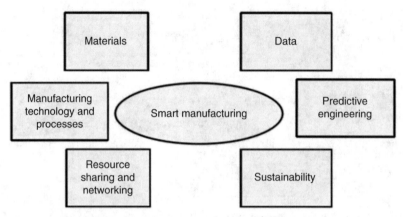

Fig. 1.4 Six pillars of smart manufacturing [8]

6. *Industry 4.0:* Smart manufacturing has continued to increase its industry penetration under different terms: smart manufacturing in the United States, Industrie 4.0 in Germany, and Smart Factory in Korea [9]. In other words, Industry 4.0 has smart manufacturing as its central element. The advancement in automation and digitization has led to Industry 4.0, which is regarded as the fourth industrial revolution. It is all about making business smarter and more automated. Industry 4.0 is revolutionizing manufacturing. It is used in manufacturing to leverage operational efficiency, refine demand forecasting, engage in predictive maintenance, offer workers boosts to safety, and more. Fig. 1.5 illustrates Industry 4.0 environment [10].

7. *Augmented Reality:* The concept of augmented reality can be applied to any technology that combines virtual as well as real information. The key objective of AR is to bring computer-generated objects into the real world and allow the user only to see it. In other words, we use AR to track the position and orientation of the user's head in order to enhance his or her perception of the world. AR is a technology that has shown promise in simplifying the operator's work during manual assembly operations in manufacturing. Due to the number of possible combinations involved, each assembly line worker must follow a unique number of operations for individual automobiles. Wearable AR can be used to help improve industrial processes. Manufacturing is one of the most promising fields where AR can be used to improve the current practices and for worker training and maintenance [11]. AR will be able to boost the skills of industrial worker. Many AR manufacturers envision AR to work like a hands-free "Internet browser" that allows workers to see real-time relevant information. AR can be employed to highlight parts on industrial equipment. It can analyze complicated machine environments and use computer vision to map out a machine's parts, like a real-time visual manual. AR has made it possible for technicians to provide remote assistance by sending customers AR-enabled devices and walking them through basic troubleshooting. Fig. 1.6 shows a typical example on how augmented reality is used in manufacturing [12].

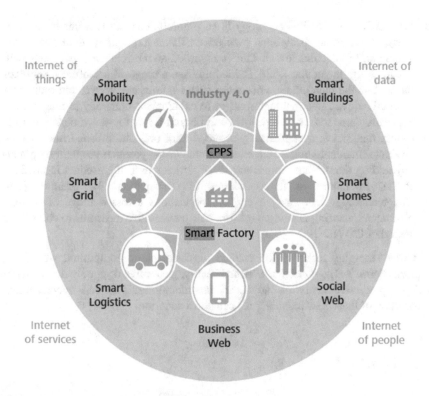

Fig. 1.5 Industry 4.0 environment [10]

Fig. 1.6 Augmented reality in manufacturing [12]

8. *Nanotechnology:* Nanotechnology is an umbrella concept that encompasses a range of scientific disciplines focused on the manipulation of matter at the atomic and molecular levels. This is regarded as the science of working with matter at the molecular scale. It encompasses a range of scientific disciplines focused on the manipulation of matter at the atomic, subatomic, and molecular levels. There are two main approaches to using nanotechnology in manufacturing (nanomanufacturing): top-down and bottom-up. In top-down nanomanufacturing, material is systematically reduced until only the finished microscopic product remains. In bottom-up nanomanufacturing, products are built by gradually adding atomic- and molecular-level particles until the product is finished, similar in nature of its operations to conventional 3D printing [13]. Nanotechnology has many everyday applications across a range of sectors. Nanomanufacturing is being used for other preventative measures to control the spread of COVID-19.

Other emerging technologies include big data, cloud computing, wearables, *drones, digital twins, blockchain,* virtual reality, generative design, materials technology, predictive analytics, advanced manufacturing, 5G, and cybersecurity. Integration of these new technologies can seem daunting.

1.3   Benefits

New technologies and innovations tend to create manufacturing jobs. Besides reducing costs and saving money, these technologies reduce waste, decrease overall production time, and enhance precision, accuracy, repeatability, efficiency, and flexibility for the manufacturing industry. From 3D printing to big data analytics to collaborative robots, the game-changing benefits of newly emerging technologies bring unprecedented transformations. They are helping manufacturers reduce human intervention, increase productivity, reduce risk, minimize downtime, and optimize market competitiveness. The new technologies will impact the manufacturing industries beyond recognition.

The emerging technologies are shaping the future of manufacturing by lowering the cost of production, improving the speed of operations, and minimizing errors. However, the manufacturing process has a long way to go before it reaches there.

Manufacturing organizations are realizing that the new technology trends have become less of an addition and more of a necessity for their business to remain competitive. Advancements of these technologies offer manufacturers opportunities to accelerate production timelines. The technologies can be used to create products that satisfy unmet needs. Integration of these trends, tools, and technologies can seem daunting, but it will result in increased efficiency. Global manufacturing companies can best capitalize on emerging technologies. They are benefitting from increase in productivity, decrease in the complexities of processes, substantial cost savings, faster production times, and the ability to provide excellent customer support.

1.4   Challenges

The twenty-first century manufacturing companies face unpredictable market changes driven by global competition, rapid introduction of new products, and constantly changing product demand. The main barrier in most countries is the lack of inventiveness for new technologies. Manufacturers are known to be slow adopters of technology. While the emerging technologies have the potential to speed up the time it takes to make a new product, the technologies are not without their drawbacks. It is challenging to implement technology solutions in a crowded vendor marketplace. Integration of the emerging technologies seems daunting.

Finding suppliers, for 3D-printed parts, for example, and keeping track of the sourcing of parts and materials and gaining trust can be difficult and time-consuming. Typically, the shop floors have old machines that still have years of production left in them, and it is not economical to replace them. Workers on the shop floor need to be more skilled than ever, and there may be no jobs for high school graduates in some manufacturing companies. Since all emerging manufacturing technologies are not created equal, further research needs to be conducted to develop specific deployment methodologies for specific technologies [14]. As the number of technologies impacting today's manufacturing increases, the bandwidth demands are intensifying.

1.5   Global Aspect of Emerging Technologies in Manufacturing

Manufacturers around the world are already employing emerging technologies to modernize their manufacturing operations to make them more efficient, cost-effective, and responsive to market demands. Researchers around the world are unlocking the potential of emerging manufacturing technologies. There is a global lack of skills base to fully exploit the emerging technologies. Resolving these challenges may take decades.

The IEEE Future Directions Committee (FDC) determines the direction of existing, new, and emerging technologies and spearheads their investigation and development. The primary objective of FDC is to incubate emerging technologies and serve as a catalyst for disseminating information about the technologies by sponsoring conferences, publications, and standards development [15].

Realizing the importance of manufacturing to their industrial future, several national governments have initiated programs to support the deployment of the new technologies in support of their domestic manufacturers [16]. Some nations like the United States, China, Singapore, and the European Union are investing heavily on the development and promotion of additive manufacturing technologies. To maximize AI adoption, nations will need privacy regimes that allow the use and reuse of data. The COVID-19 pandemic has demonstrated a global need for inexpensive, easily manufactured medical equipment to fight sudden outbreaks of disease. In the near future, one can expect reshoring to be the leading trend, as manufacturers attempt to reduce dependencies on foreign materials.

1.6 Conclusion

Manufacturing is becoming increasingly more efficient, customized, automated, and lean. As the cutting-edge technologies are being employed in industrial settings, manufacturing could be taken to unprecedented levels of production. The impact of these emerging technologies cannot be underestimated.

Manufacturing across sectors will take a long time before fully implementing these emerging technologies. Manufacturers must be willing to embrace change in order to stay ahead of competitors and win market share. They should establish long-term and short-term goals and then determine what type of technology they need to achieve their goals.

In the near future, manufacturing engineers and researchers will be in great demand to perform cost-benefit analyses, solve production issues, and operate CAD software to design and produce products. Emerging technologies are also expected to transform the type of skills required by its workforce. As manufacturers increasingly adopt new technologies, their need to hire tech-savvy workers is increasing. Unfortunately, there are not enough skilled employees to fill the open jobs. Education reform should focus on enabling existing and emerging employers to get better technical skills, particularly "twenty-first century generic skills," and produce new generation of workers. Education will liberate designers from old patterns of thinking and allow them to tap into ideation and innovation of futuristic technologies. As a Siemens executive once said, "People on the plant floor need to be much more skilled than they were in the past. There are no jobs for high school graduates at Siemens today."

We are excited and optimistic about the future and the role that digitalization, through the emerging technologies, will play to bring the concepts to reality.

References

1. M.N.O. Sadiku, T.J. Ashaolu, A. Ajayi-Majebi, and S.M. Musa, Emerging technologies in manufacturing. Int. J. Sci. Adv. 1(2) (2020)
2. M.N.O. Sadiku, S.M. Musa, O.M. Musa, Artificial intelligence in the manufacturing industry. Int. J. Adv. Sci. Res. Eng. 5(6), 108–110 (2019)
3. Future factory: How technology is transforming manufacturing (2019), https://www.cbinsights.com/research/future-factory-manufacturing-tech-trends/
4. J.B. Roca, Leaders and followers: Challenges and opportunities in the adoption of metal additive manufacturing technologies, Doctoral Dissertation, Carnegie Mellon University, Pittsburgh, 2017

5. How to integrate 3d printing into your manufacturing facility [guide] (2019), https://www.thomasnet.com/insights/how-to-integrate-3d-printing-into-your-manufacturing-facility-guide/
6. M.N.O. Sadiku, Y. Wang, S. Cui, S.M. Musa, Industrial internet of things. Int. J. Adv. Sci. Res. Eng. **3**(11), 1–4 (2017)
7. A.R. Sadeghil, C. Wachsmann, and M. Waidner, Security and privacy challenges in industrial Internet of things, in Proceedings of the 52nd Annual Design Automation Conference, 2015
8. A. Kusiak, Smart manufacturing. Int. J. Prod. Res. **56**(1–2), 508–517 (2018)
9. M.N.O. Sadiku, O.D. Olaleye, S.M. Musa, Smart manufacturing: A primer. In. J. Trend Res. Dev. **6**(6), 9–12 (2019)
10. Deloitte, Industry 4.0: Challenges and solutions for the digital transformation and use of exponential technologies (2014), https://www2.deloitte.com/content/dam/Deloitte/ch/Documents/manufacturing/ch-en-manufacturing-industry-4-0-24102014.pdf
11. M.N.O. Sadiku, C.M.M. Kotteti, S.M. Musa, Augmented reality: A primer. Int. J. Trend Res. Dev. **7**(3) (2020)
12. Augmented reality and its use case in the electronics and manufacturing industry (2020), https://arvrjourney.com/augmented-reality-and-its-future-in-the-electronics-industry-92d25eccac54
13. Manufacturing in the post-Covid world: These 5 emerging technologies could reshape the factory (2020), https://www.cbinsights.com/research/emerging-tech-reshape-manufacturing-post-covid-19/
14. D.M. Dietrich, E.A. Cudney, Methods and considerations for the development of emerging manufacturing technologies into a global aerospace supply chain. Int. J. Prod. Res. **49**(10), 2819–2831 (2011)
15. New technology connections: Future directions (n.d.), https://www.ieee.org/about/technologies.html
16. Technology and Innovation for the Future of Production: Accelerating Value Creation, World Economic Forum, Geneva, Switzerland (2017), http://www3.weforum.org/docs/WEF_White_Paper_Technology_Innovation_Future_of_Production_2017.pdf

Chapter 2
Artificial Intelligence in Manufacturing

The world's technocrats are yet to fully grasp and deploy the genius of Artificial Intelligence (AI) and the immensity of its powers to transform societies and elevate the quality of life of world citizens.

–Matthew N. O. Sadiku.

2.1 Introduction

One of the most vital industries in the world is manufacturing. Manufacturing refers to the entire product life cycle: product ideation, design, production planning, production operation, control, scheduling, distribution, supply chain, field service, and reclamation [1]. The manufacturing industry is a cornerstone of the national economy and people's livelihood.

Traditional manufacturing companies currently face several challenges such as rapid technological changes, inventory problem, shortened innovation span, short product life cycles, volatile demand, low prices, highly customized products, and ability to compete in the global markets. In today's manufacturing, technology is the determining factor for securing competitive advantage. The industry has always been receptive to adopting new technologies such as drones, industrial robots, artificial intelligence (AI), virtual reality, and Internet of Things. These technologies have come to rescue organizations such manufacturing drowned in the deluge of data, data explosion, and information overload.

Today, the manufacturing industry is using artificial intelligence and machine learning to streamline every phase of production and increase productivity.

Artificial intelligence (AI) is the cognitive science that deals with intelligent machines which are able to perform tasks heretofore only performed by human beings. It is mainly concerned with applying computers to tasks that require knowledge, perception, reasoning, understanding, and cognitive abilities. AI is potentially the algorithmic study of processes in every field of study [2]. The main objective of AI is to teach the machines to think intelligently like humans do.

Although AI is a branch of computer science, there is hardly any field which is unaffected by this technology. Common areas of applications include agriculture, business, law enforcement, oil and gas, banking and finance, education, transportation, healthcare, automobiles, entertainment, manufacturing, speech and text recognition, facial analysis, and telecommunications. Over the years, the field of AI has produced a number of tools for manufacturing. Today, industrial leaders such as Google, Microsoft, Procter & Gamble, and IBM have invested heavily on AI [3].

This chapter addresses the uses of artificial intelligence in the manufacturing industry. It begins by reviewing AI. It discusses the roles AI plays in manufacturing. It highlights the benefits and challenges of applying AI in manufacturing. It covers how AI is being applied in manufacturing around the world. It addresses the future of AI in manufacturing. The last section concludes with comments.

2.2 Overview on Artificial Intelligence

In simple terms, artificial intelligence (AI) refers to computer systems that mimic human cognitive functions. The term "artificial intelligence" (AI) was first used at a Dartmouth College conference in 1956. The main goal of AI is to enable machines to perform complex tasks that typically require human intelligence [4]. AI is now one of the most important global issues of the twenty-first century.

AI is not a single technology but a range of computational models and algorithms. AI is a collection of techniques that enables computer systems to perform tasks that would otherwise require human intelligence [5]. The major disciplines in AI include:

1. *Expert systems:* Expert system (ES) was the first successful implementation of artificial intelligence and may be regarded as a branch of AI mainly concerned with specialized knowledge-intensive domain. An expert system is a computer software that simulates the judgment and behavior of a human expert. It is also known as intelligent system or knowledge-based system. It encapsulates specialist knowledge of a particular domain of expertise leading to the making of intelligent decisions. It has a knowledge base and a set of rules that infer new facts from the knowledge. Expert systems solve problems with an inference engine that draws from a knowledge base equipped with information about a specialized domain, mainly in the form of if-then-else-if rules. It is based on expert knowledge in order to emulate human expertise in any specific field. The basic concept behind ES is that expertise (such as highly skilled medical doctor or lawyer) is transferred from a human expert to a computer system. Nonexpert users, seeking advice in the field, question the expert system to get expert's knowledge [6]. Expert systems are finding wide range of applications due to their capability to provide solutions to a variety of real-life problems. They are widely used in healthcare, business, and manufacturing. Figure 2.1 depicts an expert system [8].
2. *Fuzzy logic:* This AI discipline makes it possible to create rules for how machines respond to inputs that account for a continuum of possible conditions, rather than straightforward binary. Where each variable is either true or false (yes or no), the system needs absolute answers. However, these are not always available.

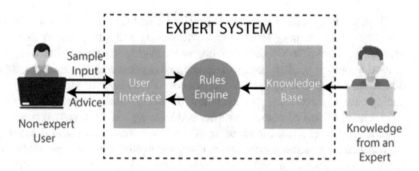

Fig. 2.1 Structure of an expert system [7]

Fuzzy logic allows variables to have a "truth value" between 0 and 1. It uses approximate human reasoning in knowledge-based systems. It was introduced in the 1960s by Lotfi Zadeh of University of California, Berkeley, known as father of fuzzy set theory. Fuzzy logic is useful in manufacturing processes as it can handle situations that cannot be adequately handled by traditional true/false logic [7].

3. *Neural networks:* This discipline of AI provides specific types of machine learning systems that consist of artificial synapses designed to imitate the structure and function of brains. An artificial neural network (ANN) is an information processing device that is inspired by the way the brain processes information. They were originally developed to mimic the learning process of the human brain. The idea of ANNs was inspired by the structure of the human brain and by the synthesis, simulation, extraction, codification, and processing of what the brain can do. They may be regarded as a sort of parallel processor designed to imitate the way the brain accomplishes tasks. They are made up of artificial neurons that take in multiple inputs and produce a single output. In other words, the purpose of a neuron is to collect inputs and to generate outputs. The network observes and learns as the synapses transmit data to one another, processing information as it passes through multiple layers [9].

4. *Machine learning:* Machine learning (ML) is essentially the study of computer algorithms that improve automatically through experience and iterative pattern associations, recognitions, looping, and uses. It is the field that focuses on how computers learn from data. This includes a broad range of algorithms and statistical models that make it possible for systems to find patterns, draw inferences, and learn to perform tasks without specific instructions. Machine learning is a process that involves the application of AI to automatically perform a specific task without explicitly programming it. Learning algorithms work on the assumption that strategies, algorithms, and inferences that worked well in the past are likely to work well in the future. ML techniques may result in data insights that increase production efficiency. Using ML can save time for practitioners and provide unbiased, repeatable results. Today, artificial intelligence is narrow and mainly based on machine learning.

There are two types of learning: supervised learning and unsupervised learning. Supervised learning focuses on classification and prediction. It involves building a statistical model for predicting or estimating an outcome based on one or more inputs. It is often used to estimate risk. Supervised ML is where algorithms are given training data. In unsupervised learning, we are interested in finding naturally occurring patterns within the data. Unlike supervised learning, there is no predicted outcome. Unsupervised learning looks for internal structure in the data. Unsupervised learning algorithms are common in neural network models. Machine learning techniques have been applied in the analysis of data in various fields including medicine, finance, business, education, advertising, cybersecurity, and manufacturing [10]. Manufacturers are turning to machine learning to improve the end-to-end performance of their operations.

5. *Deep learning*: This is a form of machine learning based on artificial neural networks. Deep learning may be regarded as a software technology designed by programmers to teach computers to do what humans have been doing: learning by example or learning from data. Deep learning (DL) architectures are able to process hierarchies of increasingly abstract features, making them especially useful for purposes like speech and image recognition and natural language processing. Deep learning networks can deal with complex nonlinear problems. It extracts complex features from high-dimensional data and applies them to develop a model that relates inputs to outputs. The most common form of deep learning architectures is multilayer neural networks. Deep learning has many advantages over shallow learning. Due to this, deep learning networks have received much attention as they can deal with more complex nonlinear problems.

   Recently, companies such as IBM, Microsoft, Google, Apple, and Baidu have invested and developed deep learning. They have taken advantage of their massive data and large computational power to deploy deep learning on a large scale. DL has several applications in the manufacturing domain including sales forecasting and advanced analytics. Although DL has achieved some success and found applications in various fields, it is still in its infancy [11].

6. *Natural Language Processors*: For AI to be useful to us humans, it needs to be able to communicate with us in our language. Language is crucial around the world in communication, entertainment, media, culture, drama, movie, and economy. Computer programs can translate or interpret language as it is spoken by normal people. Natural language processing (NLP) refers to the field of study that focuses on the interactions between human language and computers. It is a computational approach to text analysis. It involves the study of mathematical and computational modeling of various aspects of language. It is an interdisciplinary field involving computer science, linguistics, logic, and psychology.

   NLP is important because of the major role language such as English plays in human intelligence and because of the wealth of potential applications. Applications of NLP include interfaces to expert systems and database query systems, machine translation, text generation, story understanding, automatic speech recognition, and computer-aided instruction. It also has great potential in healthcare, mobile technology, cloud computing, virtual reality, election, social work, and social networking [12].

7. *Robots*: AI is heavily used in robots, which are computer-based programmable machines that have physical manipulators and sensors. Sensors can monitor temperature, humidity, pressure, and time, record data, and make critical decisions in some cases. Robots have moved from science fiction to your local hospital. In jobs with repetitive and monotonous functions, they might even completely replace humans. Robotics and autonomous systems are regarded as the fourth industrial revolution.

   Robotics is a branch of engineering and computer science that involves the conception, design, manufacture, and operation of robots. A robot functions as an intelligent machine, meaning that it can be programmed to take actions or make choices based on input from sensors. It involves using electronics, computer science, artificial intelligence, mechatronics, and bioengineering. Robots are applied in many fields including agriculture, education, manufacturing, entertainment, medicine, space exploration, undersea exploration, sex, power grid, agriculture, construction, meat processing, household, mining, aerospace, electronics, and automotive. For example, robot police with facial recognition technology have started to patrol the streets in China. Future robots will operate in highly networked environments where they will communicate with other systems such as industrial control systems and cloud services [13].

These AI tools are illustrated in Fig. 2.2 [15]. Each AI tool has its own advantages. Using a combination of these AI technologies and models, rather than a single model, significant organizational, operational, and productivity advantages are unleashed. AI systems are designed to make decisions using real-time data. They have the ability to learn and adapt as they make decisions. Artificial intelligence is no longer just an academic field; machine learning and deep learning are becoming mainstream technologies that manufacturing companies can harness.

AI-driven systems can discover trends, reveal inefficiencies, and predict future outcomes. These characteristics enable informed decision-making and make AI to be potentially beneficial for many industries. Several companies have realized the value of AI technology. By building precise models, a company has a better chance of identifying profitable opportunities and avoiding risks. Machine learning techniques are being applied in several sectors including government, healthcare, finance, retail, transportation, oil and gas, and manufacturing. Over the years, the manufacturing industry has exploited the use of artificial intelligence (AI), which is the intelligence exhibited by machines. The factors that make AI best suitable for manufacturing include [16]: virtual reality, automation, intelligence, prediction, and better products.

2.3   AI in Manufacturing

Manufacturing is regarded as a prime generator of wealth and is critical in establishing a sound basis for economic growth. Globalization has radically changed the manufacturing process in recent years. The manufacturing industry has become

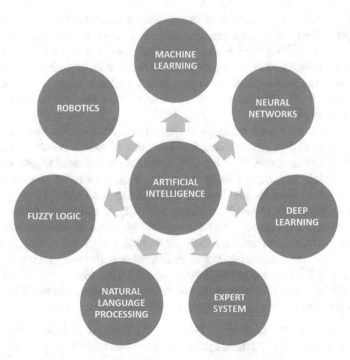

Fig. 2.2 Branches of artificial intelligence [14]

more competitive due to globalization. To survive and indeed thrive, in the global competitive market, manufacturing enterprises should reduce the product cost and increase the productivity while not sacrificing performance, quality, durability, reliability, availability, serviceability, aesthetics, feature, and conformance to standards and/or specifications.

Manufacturing consists of a set of processes, machines, and factories where raw materials are transformed into products. The search for new ways to increase efficiency and productivity has always been at the heart of manufacturing. To meet these challenges, it is essential to utilize all means available. One area, which saw fast pace of developments in terms of not only promising results but also usability, is artificial intelligence. AI is one of the many technologies currently impacting manufacturing [17].

Manufacturing efficiency and maintaining product quality remain top priorities in some industries such as IC industry, where process and equipment reliability directly influence cost, throughput, and yield [14]. Recent developments have enabled AI to cross into the mainstream: cloud computing, big data, machine learning, and Internet of Things (IoT).

The manufacturing industry is now moving into the fourth industrial revolution:

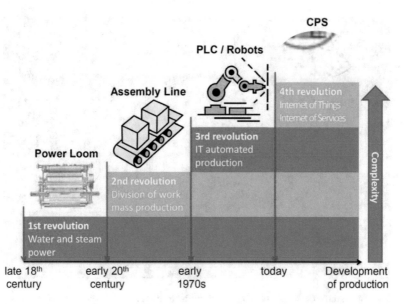

Fig. 2.3 Four industrial revolutions [18]

- The first revolution introduced mechanization through water and steam power.
- The second revolution ushered in mass production and assembly lines using electricity.
- The third industrial revolution introduced the digital era, comprised of computers and automation.
- The fourth industrial revolution, referred to as "Industry 4.0," is happening now.

    These revolutions are illustrated in Fig. 2.3 [19].

2.4   Applications of AI in Manufacturing

The applications of artificial intelligence are everywhere, and its uses are constantly evolving. AI promises to be a game-changer at every facet of manufacturing and might help in solving some of its so far unsolvable problems. The techniques of AI in general, and machine learning (ML) in particular, have been a source of inspiration for researchers in the manufacturing industry. We now present some applications of AI in manufacturing:

- *Robotics:* Perhaps the most promising way is applying machine learning algorithms to the traditional manufacturing system. Even with advanced manufacturing techniques, using humans to spot defects and errors is very limiting. Today, robotics has revolutionized manufacturing, allowing for greater output from fewer workers, while AI is just beginning to live up to its full potential. A robot

Fig. 2.4 An example of the use of robots in manufacturing [20]

is a mechatronic device that is designed and programmed to perform some specific tasks. Robots typically use sensor data to make decisions. Robots are used in diverse manufacturing applications such as pick and place, sorting, assembly, palletizing, jigging, fixturing, harnessing, positioning, painting, welding, storage, and retrieval. AI coupled with computer vision techniques allows autonomous robots to complete these tasks. A typical example of the use of robots in manufacturing is shown in Fig. 2.4 [21].

- *Automation:* Today, the manufacturing industry is experiencing major changes due to aggressive global competition, increasing customer expectations, and expanding choice of materials. To cope with increasing demands of today's dynamic and competitive market, existing manufacturing processes and tasks are automated. In the hope of decreasing costs and increasing quality and the push for higher volumes of output with lower investments, manufacturing industry is making efforts to achieve a higher degree of automation. Manufacturers are attempting to have their manufacturing systems partially or fully automated in order to reduce labor costs, increase operational efficiencies, and increase production rates. For years, machine builders in the automation field of assembly equipment have been focusing on the mechanics of machines. Fig. 2.5 illustrates automation in manufacturing [23].
- *Automotive Industry:* The automotive industry covers a wide range of vehicles. The industry faces severe challenges at the moment due to the trend toward a connected, autonomous, and green individual mobility, which requires high investments. Deep learning has many potential applications in the automotive industry during development, manufacturing, and sales. It is also useful in advanced driving assistance systems, autonomous driving, and advanced detec-

Fig. 2.5 Automation in manufacturing [22]

tion controls. Although there are many automotive suppliers all over the world, they essentially follow the same procedure in their production [24, 25]. AI can help insurance companies in assessing risk as well as helping customers in filing their insurance claims. Predicting future trends can make a big difference in the automotive insurance sector.

- *Industry 4.0*: The manufacturing industry today is experiencing a never seen increase in available data. These data that can be obtained from manufacturing system consists of man, machine, material, and method data (i.e., 4 M data). Different names are used for this phenomenon, e.g., Industrie 4.0 (Germany), smart manufacturing (USA), and smart factory (South Korea). Industry 4.0 refers to the current trend of automation and employment of AI technologies in manufacturing. The major applications of Industry 4.0 are smart factory, smart manufacturing, smart product, and smart city. Those who promote Industry 4.0 claim that it will affect many areas such as services and business models, productivity, machine safety, product lifecycles, and industry value chain [18, 26]. The terms "smart factory," "smart manufacturing," "Industry 4.0" and "factory of the future" represent keywords to denote what industrial production will look like in the future. They refer to a factory with a manufacturing solution that provides a flexible, adaptive, and efficient production. As shown in Fig. 2.6, the implementation of the smart factory is enabled by several technologies such as artificial intelligence, industrial robots, embedded systems, RFID, cloud computing, big data, augment reality, and the Internet of Things (IoT) or industrial Internet of Things (IIoT) [20, 28].

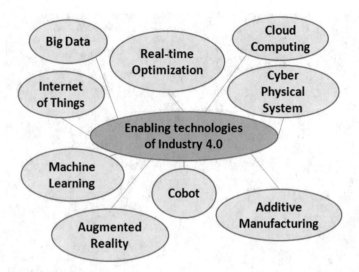

Fig. 2.6 Enabling technologies of Industry 4.0 [27]

- *Predictive Manufacturing:* Nearly all engineering systems are subject to failures
  due to deterioration or misuse. Instead of fixing things when failure happens, it
  is better to predict problems before they occur. Manufacturers have faced a compelling need for the development of predictive models that predict mechanical
  failures or predict when maintenance is due. The central objective of applying AI
  and machine learning techniques to manufacturing systems is to accomplish the
  predictive manufacturing. This is a manufacturing system that has self-awareness
  and can detect current conditions, diagnose a problem, and make a decision. This
  may also be regarded as predictive maintenance, which is able to predict disrup-
  tions to the production line in advance of that disruption [22]. Data collected
  from sensors can be used to predict the failures and do preventive maintenance
  which is a lot cheaper than fixing problem situations after failure. Predictive
  manufacturing is required in order to realize the smart factory. The main benefit
  of AI-based predictive maintenance is accuracy and promptness.
- *Semiconductor Manufacturing*: This is a complex process that requires monitoring a great number of parameters from production to the packaging. VLSI technology has advanced to the stage where the automation of semiconductor
  manufacturing has become imminent. Expert system technology can be applied
  to monitor various processes and equipment for the IC manufacturing environment. Machine learning is a useful tool in the automation of the IC manufacturing process. It also serves as an aid to engineers in interpreting and assimilating
  experimental results [29]. Within the semiconductor wafer manufacturing process, tight quality control is extremely important. Also, the volume of data in
  semiconductor manufacturing makes the data analysis difficult and time-
  consuming. Most semiconductor manufacturers keep a vast database to record
  every parameter in the manufacturing process. This online availability of vast

amounts of production data brings the need for automatic analysis tools, to which machine learning techniques are good candidates [30].

- *Supply Chain Management:* AI can be used to collect and monitor data along the supply chain and then manage inventory, predict future demand, and detect inefficiencies. Artificial intelligence and machine learning make supply chain management not only automated but cognitive. The cognitive supply chain ensures data privacy and prevents hacking. Supply chain management systems based on AI algorithms can automatically analyze such data, define optimal solutions, and make data-driven decisions. Supply chain management is an example of how AI and big data can be used to the benefit of manufacturers.
- *Intelligent Manufacturing:* The continuous evolution of smart cities, intelligent transportation, intelligent robots, smartphones, smart communities, smart grid, etc. provides a market demand and driving force for both AI technologies and applications. This has led to a new manufacturing model known as intelligent or smart manufacturing. AI is applied in intelligent manufacturing through the intelligent manufacturing system. The intelligent manufacturing system is characterized by autonomous intelligent sensing, intelligent equipment, intelligent robots, new-generation intelligent network devices, networking, collaboration, learning, analysis, cognition, knowledge-based intelligence design, prediction, decision-making, and on-demand services at any time and any place. Intelligent manufacturing system is used to facilitate production and provide a high efficiency, high-quality, cost-effective, and environment-friendly service to users [31]. Figure 2.7 shows the components of an intelligent manufacturing system [33].
- *Predictive Maintenance:* In predictive maintenance, data is collected overtime to monitor and find patterns to predict failures. Although predictive maintenance is applied in many industries, it thrives in manufacturing. AI in manufacturing has come a long way with technologies like predictive maintenance. Machine learn-

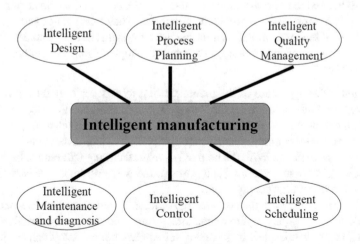

Fig. 2.7   Components of an intelligent manufacturing system [32]

ing can be used to predict recommended schedules of preventative maintenance in order to improve worker safety, reduce costs, and achieve sustainability goals. The adoption of machine learning and deep learning in manufacturing will only improve predictive maintenance. Deep learning can aid in the predictive maintenance of complex machinery and connected systems.

- *Manufacturing Site Selection:* The need for the development of the prototype expert system that enables optimal and efficient site selection comes from state or local planners, who must identify possible sites when asked by potential foreign manufacturing investors, who are looking for suitable locations for production plants. Site selection is one of the most crucial decisions that industrial management has to make in considering establishment of a new plant.

Artificial intelligence is also used in additive manufacturing, digital manufacturing, smart manufacturing, cellular manufacturing, flexible manufacturing, adaptive manufacturing, green manufacturing, manufacturing healthiness, furniture manufacturing, and "service part, maintenance, and reusable part" remanufacturing. AI has also found a niche in product design, generative design, process control, fluid power industry, tire industry, aerospace industry, food and beverage industry, shipbuilding industry, textile industry, steel industry, apparel industry, and construction industry.

2.5 Benefits

Artificial intelligence can help address some issues facing manufacturing, improve quality control, shorten design cycles, remove supply-chain bottlenecks, reduce materials and energy waste, and improve production yields. AI tools are helping manufacturers find new business models, improve their product quality, increase productivity, reduce chronic labor shortage, and optimize manufacturing operations, while finding new ways to retain employees. They can enhance and extend the capabilities of humans. Manufacturers can use AI systems to design smart products, forecast demand, ensure quality, and manage supply chain risk. Other benefits include [34]:

- *Robotics:* A large share of the manufacturing jobs is performed by robots. AI-powered robots can interpret CAD models, which eliminate the need to program their movements and processes. Industrial robots perform what they were programmed to do, but they cannot make complex decisions as humans.
- *Higher Productivity:* With the help of AI, manufacturing will realize higher productivity, increase efficiency, and improve the way customers are served. AI is poised to impact the supply chain.
- *Better Decisions:* With the adoption of AI, manufacturers are able to make rapid, data-driven decisions, optimize manufacturing processes, and minimize operational costs. While machines cannot yet surpass human intelligence, they can

outperform humans in relation to speed, accuracy, and ability to search, sort, transform, interpret, and analyze vast amounts of information.

- *Price Forecasts:* Knowing the prices of resources is also necessary for companies when their products are ready to leave the factory. The system is able to provide accurate price recommendations.
- *Customer Service*: AI solutions can analyze the behaviors of customers to identify patterns and predict future outcomes. There are a number of AI tools that can improve customer service.

2.6 Challenges

The field of AI is broad and even confusing, presenting a challenge and hindering wide applications. Manufacturing is a high-cost venture for most businesses, and it requires considerable capital investments. It is possible for AI to develop a will of its own that may be in conflict with that of humans, especially when under the influence of clandestine or cryptic cyberattacks. The increasing use of connected technologies makes the manufacturing system vulnerable to cyberattacks [27]. Other challenges include [34]:

- *Complexity*: A major challenge is the ongoing trend of the manufacturing domain to becoming more complex and dynamic. Due to this challenge, modern manufacturing technology is interdisciplinary in nature.
- *Common Fear:* There is the common fear of human jobs being lost to AI. Mass unemployment will be a social challenge. This will require retraining human workers for new careers. With automation and increased deployment of AI technologies, it is conceivable that machines may outperform human workers in many areas of human technology-based endeavors.
- *Environmental Impact*: The manufacture of a variety of products may continue to damage the environment when there is disregard for design for environment and design for recyclability.
- *Shortage in Workforce:* Manufacturing companies will feel the challenge of a decreasing talent pool. There are not enough skilled people to perform the AI-related jobs of the future. There continues to be a shortage of AI-based technological know-how within the manufacturing workforce.
- *Liability:* AI raises important legal, ethical, and public policy questions. AI technology implementers or AI systems integrators will have to make potentially life-or-death decisions. How should a self-driving car decide between crashing itself and potentially killing its passengers? How do we assign liability for when AI-enabled and/or powered systems make mistakes?

2.7   Global AI in Manufacturing

Artificial intelligence is gradually being implemented in almost every nation. In recent years, the global market has witnessed significant progress in technologies, industry, and applications in terms of manufacturing in general and intelligent manufacturing in particular. With the inroading of artificial intelligence (AI) and Internet of Things (IoT) into the manufacturing sector, nations are inevitably faced with a number of policy concerns which need to be addressed at various levels keeping in mind the socioeconomic factors that influence policy making in that particular nation. For example, the United States and Germany have spared no effort in demonstrating and promoting their development strategy to transform and upgrade their manufacturing industries.

The Global Manufacturing and Industrialization Summit (GMIS) was established in 2015 to build bridges between manufacturers, governments and NGOs, technologists, and investors so that they can harness the transformative power of the Fourth Industrial Revolution. Here, we consider how AI is being applied in manufacturing in different countries [18, 35].

- *United States:* The leading manufacturing industries in the United States include steel, automobiles, chemicals, food processing, consumer goods, aerospace, and mining. Manufacturing technology in the United States has accomplished tremendous outcomes. The Boeing Company and General Electric have identified challenges to the incorporation of AI in the manufacturing sector. The Manufacturing Institute and Deloitte estimate that the United States alone may have as many as 2.4 million manufacturing jobs to fill between now and 2028. Americans have not yet grappled with just how profoundly the artificial intelligence (AI) revolution will impact our economy, national security, and welfare. We take seriously China's ambition to surpass the United States as the world's AI leader within a decade. The US Government is accelerating its commitment in AI techniques due to the fact that the manufacturing sector is ripe for accelerated growth [36].
- *China:* China is well known as the "manufacturing kingdom." Manufacturing industry in China is facing a critical and historical moment in terms of the transformation of the manufacturing giant to manufacturing power and from "made in China" to "created in China." The industry is in a phase of imbalanced development among different regions, industries, and enterprises, where mechanization, electrification, automation, and digitization coexist. The Chinese government has proposed strategic plans of "made in China 2025." It sets the principles for becoming the second tier of the powerhouse in manufacturing: innovation-driven, quality first, environment-friendly development, structural optimization, and talent-centric innovations. The plan also includes the following nine missions: (1) increasing innovative capability in national manufacturing, (2) promoting the deep fusion of information and industrialization, (3) strengthening the basic industry capacity, (4) booming the quality brand-building, (5) popularizing environment-friendly manufacturing, (6) advancing breakthroughs in key areas,

(7) pushing forward further the structural adjustment to the manufacturing industry, (8) advancing services for manufacturing and production, and (9) increasing international involvement in manufacturing. The manufacturing industry in China is faced with imbalanced development among different regions, industries, and enterprises. The AI technology and its deployment are gradually changing the economy of the world in general and of China in particular [37].

- *Germany:* Germany mainly focuses on the research of underlying technologies for manufacturers, such as intelligent sensing, wireless sensor networks, and CPSs. Sensors will gather data from the engine during its flight and transmit it to the ground for smart analysis. The Germany Amberg factory is a model of an intelligent plant of the Siemens Company. At Amberg, the real factory is operated together with the virtual factory, and the real factory data and production environments are reflected by the virtual factory.
- *India:* The impact of AI on the manufacturing industry in India often depicts a picture of a stagnant job growth and even job loss. AI is disrupting traditional business models in the IT sector, the automotive industry, and other manufacturing industries. However, smart manufacturing are picking up in India. The Indian Institute of Science is developing a smart factory. India, unlike its G20 counterparts, is yet to fully tap the available opportunities that AI presents.

2.8   Future of AI in Manufacturing

Global competition and continuously changing customer requirements are dictating increasing changes in manufacturing environments. AI in manufacturing is a game-changer. AI is already transforming manufacturing in several ways and also changing the way we design products. Today, manufacturing has a strong association with AI, especially in the use of automation. AI technologies in the manufacturing industry are expected to grow over the next several years. There is an excellent future for use of AI in manufacturing due to the potentially high value-added prospects when applied successfully [36]. The future will witness a deep fusion of new AI technologies with Internet technologies. AI will have a fundamental impact on the global labor market in the next few years.

Today's technologies are able to predict design successes, product performances, and equipment failures, providing key information that helps manufacturers continuously improve. AI algorithms formulate estimations of market demands by looking for patterns linking location, socioeconomic and macroeconomic factors, weather patterns, political status, consumer behavior, and more. This information is invaluable to manufacturers as it allows them to optimize staffing, inventory control, energy consumption, and the supply of raw materials.

Some experts in the field of manufacturing are already making predictions on what is coming and the trends driving AI advancement. Figure 2.8 illustrates the future of AI in manufacturing [37]. Here are three exciting trends to watch out for in the coming years:

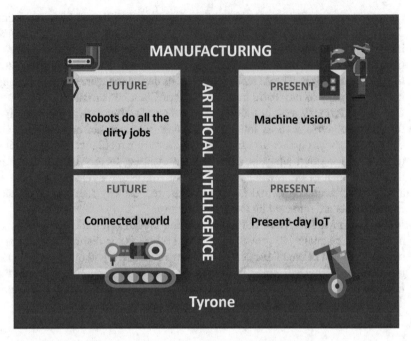

Fig. 2.8 The future of AI in manufacturing. (Modified [37])

- *Advanced Manufacturing:* This is a family of activities that depend on the use and coordination of information, automation, computation, software, sensing, and networking. It uses cutting edge materials and emerging technologies such as nanotechnology and industrial IoT. In order to capture the potential of advanced manufacturing and create new advantages, we must collaborate to bring these emerging technologies into the next generation. With artificial intelligence, advanced manufacturing can become more autonomous, efficient, and profitable [38].
- *Smart Manufacturing:* Smart manufacturing (SM) is a term coined by several agencies in the United States and increasingly used globally. Other initiatives such as Industry 4.0, cyber-physical production systems, smart factory, intelligent manufacturing, and advanced manufacturing are frequently used synonymously with smart manufacturing. SM is characterized by the integration of new-generation information communication technologies (ICT) in support of the manufacturing industry. Smart manufacturing, which is the fourth revolution in the manufacturing sector, has its own identity captured in the following six pillars [39]: (1) data, (2) materials, (3) sustainability, (4) predictive engineering, (5) resource sharing and networking, (6) manufacturing technology and processes. Smart manufacturing will transform how products are designed, fabricated, operated, and used. Globalization and competitiveness are forcing companies to rethink and innovate their production process [32].

- *Additive Manufacturing:* Additive manufacturing (AM) (or 3D printing or layered manufacturing or fused deposition manufacturing or laminated object manufacturing) refers to a process by which 3D CAD data is used to construct an object in layers by depositing material. The promise of AM is the use of less material and to provide faster cycle time due to rapid prototyping. AM technologies typically include a computer, 3D modeling software (Computer Aided Design or CAD), a machine, and material. The AM machine reads in data from the CAD data file and develops successive layers of the material in a layer-upon-layer manner to fabricate the desired 3D object. Additive manufacturing allows one to produce the objects without machining, lathing, milling, grinding, boring, casting, or welding. It will take manufacturing to the next level. It will lead to home manufacturing [40].
- *Explainable AI:* This bridges the transparency and trust between AI technology and its consumers. Explainable AI in manufacturing improves efficiency, workplace safety, and customer satisfaction using automation. It can save a manufacturing enterprise money and time by reducing and predicting machine breakdowns. To implement Explainable AI in our AI system requires obeying four principles: (1) Explanation: The system will provide a reason for each decision. The predictive maintenance system will tell when failure will occur and the need for maintenance or equipment replacement. (2) Meaningful: The explanation that the system provides should be understandable to the targeted recipient. The developer needs to know how the system reaches its decision so that they can trust the system. (3) Explanation Accuracy: The explanations provided by the system should be accurate. If it generates the wrong output, then it is of no use. (4) Knowledge Limits: This prevents the system from giving an unjust and fallacious result. Hence, users can assure that the system will never mislead [41].

2.9 Conclusion

Artificial intelligence is changing the world and impacting our lives. AI deals with computational systems that imitate the intelligent behavior of experts. Considerable progress has been made in the application of AI in several aspects of manufacturing. For manufacturers considering introducing AI into their processes, it is important to define their goal from the outset. Manufacturers need to ask themselves where they want to use AI in the near, mid-, and long term.

AI is both a powerful source of game-changing technology disruption and a tool to gain highly leveraged competitive advantage.

AI technologies are still rapidly evolving, often causing uncertainty for companies around what their future IT architecture should look like. For manufacturers, AI promises to be a game-changer at every level of the value chain. AI technology is now making its way into manufacturing and could hold the key to transforming factories of the near future.

Without doubt artificial intelligence holds the key to future growth and success in manufacturing. The combination of manufacturing technology with information communication technology and smart technology is enabling a game-changing transformation in terms of manufacturing models and approaches. AI is poised to change the way we manufacture products and process materials forever. More information about AI in manufacturing can be found in the book in [42–47] and the following related journals:

- *Artificial Intelligence*
- *Applied Artificial Intelligence*
- *Artificial Intelligence for Engineering Design, Analysis and Manufacturing*
- *Journal of Artificial Intelligence and Consciousness*
- *Journal of Intelligent Manufacturing*
- *Journal of Artificial General Intelligence*
- *Journal of Experimental & Theoretical Artificial Intelligence*
- *AI Magazine*
- *AI & Society*
- *Artificial Intelligence in Agriculture*
- *IEEE Transactions on Artificial Intelligence*
- *IEEE Journal on Robotics and Automation*
- *Manufacturing Letters*

References

1. M.S. Fox, Industrial applications of artificial intelligence. Robotics **2**, 301–311 (1986)
2. M.N.O. Sadiku, Artificial intelligence. IEEE Potentials **8**, 35–32 (1989)
3. M.N.O. Sadiku, S.M. Musa, O.M. Musa, Artificial intelligence in the manufacturing industry. Int. J. Adv. Sci. Res. Eng. **5**(6), 108–110 (2019)
4. H. Chen, L. Li, Y. Chen, Explore success factors that impact artificial intelligence adoption on telecom industry in China. J. Manage. Anal. (2020)
5. S. Greengard, What is artificial intelligence? (2019), https://www.datamation.com/artificial-intelligence/what-is-artificial-intelligence.html
6. M.N.O. Sadiku, Y. Wang, S. Cui, S.M. Musa, Expert systems: A primer. Int. J. Adv. Res. Comput. Sci. Softw. Eng. **8**(6), 59–62 (2018)
7. G. Singh, A. Mishra, D. Sagar, An overview of artificial intelligence. SBIT J. Sci. Technol. **2**(1) (2003)
8. What is an Expert System? (n.d.), https://www.javatpoint.com/expert-systems-in-artificial-intelligence
9. M.N.O. Sadiku, S.M. Musa, O.S. Musa, Neural networks in the chemical industry. Invent. J. Res. Technol. Eng. Manage. **1**(12), 25–27 (2017)
10. M.N.O. Sadiku, S.M. Musa, O.S. Musa, Machine learning. Int. Res. J. Adv. Eng. Sci. **2**(4), 79–81 (2017)
11. M.N.O. Sadiku, M. Tembely, S.M. Musa, Deep learning. Int. Res. J. Adv. Eng. Sci. **2**(1), 77–78 (2017)
12. M.N.O. Sadiku, Y. Zhou, S.M. Musa, Natural language processing. Int. J. Adv. Sci. Res. Eng. **4**(5), 68–70 (2018)

13. M.N.O. Sadiku, S. Alam, S.M. Musa, Intelligent robotics and applications. Int. J. Trends Res. Dev. **5**(1), 101–103 (2018)
14. G.S. May, T.S. Kim, G. Triplett, I. Yun, Artificial intelligence in semiconductor manufacturing, in *Wiley Encyclopedia of Electrical and Electronics Engineering*, ed. by J. Webster, (Wiley, 2007)
15. How does artificial intelligence work? (n.d.), https://www.innoplexus.com/blog/how-artificial-intelligence-works/
16. T. Nicholas, The future of artificial intelligence in manufacturing industries (2018), https://www.intellectyx.com/blog/the-future-of-artificial-intelligence-in-manufacturing-industries/
17. M.N.O. Sadiku, Y.P. Akhare, S.M. Musa, Machine learning in manufacturing: A brief survey. Int. J. Trend Res. Dev. **6**(6) (2012)
18. M.N.O. Sadiku, S.M. Musa, O.M. Musa, The essence of industry 4.0. Invent. J. Res. Technol. Eng. Manage. **2**(9), 64–67 (2018)
19. P.P. Khargonekar, Artificial intelligence, automation, and manufacturing (2018), https://cpb-us-e2.wpmucdn.com/faculty.sites.uci.edu/dist/8/644/files/2018/06/Smart_MFG_CNMI.pdf
20. A. Suresh, R. Udendhran, and M. Balamurugan, Integrating IoT and machine learning – The driving force of industry 4.0, in *Internet of Things for Industry 4.0*, ed. by G. Kanagachidambaresan et al. (Springer, 2020)
21. B. Ramsey, How machine learning is poised to revolutionize manufacturing (n.d.), https://medium.com/supplyframe-hardware/how-machine-learning-is-poised-to-revolutionize-manufacturing-7e72b4ba8e5f
22. J.H. Han, R.Kim, and S.Y. Chi, Applications of machine learning algorithms to predictive manufacturing: Trends and application of tool wear compensation parameter recommendation, in Proceedings of the 2015 International Conference on Big Data Applications and Services (Jeju Island, 2015), pp. 51–57
23. Our vision is for AI-delivered insights tied into automation – Kraft Heinz on using artificial intelligence in manufacturing (2020), https://www.just-food.com/interview/our-vision-is-for-ai-delivered-insights-tied-into-automation-kraft-heinz-on-using-artificial-intelligence-in-manufacturing
24. A. Luckow et al., Deep learning in the automotive industry: Applications and tools, in Proceedings of 2016 IEEE International Conference on Big Data (Big Data), 2016
25. Q. Demlehner, D.I. Schoemer, S. Laumer, How can artificial intelligence enhance car manufacturing? A Delphi study-based identification and assessment of general use cases. Int. J. Inf. Manag. **58**, 102317 (2021)
26. T. Wuest et al., Machine learning in manufacturing: Advantages, challenges, and applications. Prod. Manuf. Res. **4**(1), 23–45 (2016)
27. G. Jujjavarapu, E. Hickok, and A. Sinha, AI and the manufacturing and services Industry in India (n.d.), \https://cis-india.org/internet-governance/files/AIManufacturingandServices_Report_02.pdf
28. J. Lee et al., Industrial artificial intelligence for industry 4.0-based manufacturing systems. Manuf. Lett. **18**, 20 (2018)
29. K.B. Irani et al., Application of machine learning techniques to semiconductor manufacturing, in Proc. SPIE 1293, Applications of Artificial Intelligence VIII, 1990
30. C.K. Shin, S.C. Park, A machine learning approach to yield management in semiconductor manufacturing. Int. J. Prod. Res. **38**(17), 4261–4271 (2000)
31. B. Li et al., Applications of artificial intelligence in intelligent manufacturing: A review. Front. Inf. Technol. Electron. Eng. **18**, 86–96 (2017)
32. M.N.O. Sadiku, O.D. Olaleye, S.M. Musa, Smart manufacturing: A primer. In. J. Trend Res. Dev. **6**(6), 9–12 (2019)
33. F. Meziane et al., Intelligent systems in manufacturing: Current developments and future prospects (n.d.), https://www.researchgate.net/publication/242176196_Intelligent_systems_in_manufacturing_Current_developments_and_future_prospects/link/02bfe5110ffd9b3797000000/download

34. K. Polachowska, 10 use cases of AI in manufacturing (2019), https://neoteric.eu/blog/10-use-cases-of-ai-in-manufacturing/
35. T.M. Knasel, Artificial intelligence in manufacturing: Forecasts for the use of artificial intelligence in the USA. Robotics **2**, 357–362 (1986)
36. Y. Liu, Research on the construction method of mechanical manufacturing system based on artificial intelligence technology, in Proceedings of 2020 5th International Conference on Smart Grid and Electrical Automation, 2020, pp. 149–152
37. Artificial intelligence: Future & present in manufacturing industry (n.d.), https://blog.tyrone-systems.com/artificial-intelligence-future-and-present-in-manufacturing-industry
38. Machine learning & artificial intelligence in advanced manufacturing (n.d.), https://www.rit.edu/gis/coe/sites/rit.edu.gis.coe/files/docs/news-events/Machine%20Learning%20Blog.pdf
39. A. Kusiak, Smart manufacturing. Int. J. Prod. Res. **56**(1–2), 508–517 (2018)
40. K.S. Prakasha, T. Nancharaihb, V.V.S. Rao, Additive manufacturing techniques in manufacturing: An overview. Mater. Today: Proc. **5**, 3873–3882 (2018)
41. J.K. Gill, Explainable AI in manufacturing industry (2021), https://www.akira.ai/blog/ai-in-manufacturing-industry/
42. J.P. Davim, *Artificial Intelligence in Manufacturing Research* (Nova, 2010)
43. R.K. Miller, T.C. Walker, *Artificial Intelligence Applications in Manufacturing*, 2nd edn. (Prentice Hall, Hoboken, 1982)
44. A.F. Famili, D.S. Nau, S.H. Kim (eds.), *Artificial Intelligence Applications in Manufacturing* (MIT Press, Cambridge, 1992)
45. G. Shaw, *The Future Computed: AI & Manufacturing* (The Future Computed Series) (Independently Published, 2019)
46. J. Krakauer, *Smart Manufacturing with Artificial Intelligence* (Manufacturing update series) (Society of Manufacturing Engineers, 1987)
47. S.K. Mishra et al. (eds.), *AI in Manufacturing and Green Technology: Methods and Applications* (CRC Press, Boca Raton, 2020)

Chapter 3
Robotic Automation in Manufacturing

Advances in automation, artificial intelligence and robotics, while increasing productivity, will also cause major upheavals to the workforce.

–John Hickenlooper.

3.1 Introduction

The manufacturing sector represents the largest private industry in the United States. It is well-known to be a major contributor in both economic boom times and economic recessions. To boost productivity, manufacturing companies turn to advanced technology such as robotic automation. Automation refers to any technology that reduces the need for human assistance. It essentially describes mechanization, machines replacing human labor, or human decision-making. It can have far-reaching consequences in manufacturing.

Robotic automation has become one of the key areas for modern manufacturing systems. It is applicable to virtually any industry imaginable, just as automotive, electronics, food, and manufacturing. It is used in manufacturing to change the industrial landscape by increasing productivity, repeatability, and precision while protecting employees from unsafe working environments [1]. Automation can assist the manufacturing industry by reducing labor costs substantially; increasing quantity or generating more output for a given period of time; enhancing accuracy, repeatability, reliability, and safety; improving product quality; and mitigating risk.

Several economists believe employment in routine occupations has declined and automation is a wave of technological change that could lead to structural shifts in the labor market resulting in "job polarization." One of the most important application areas for deploying automation is manufacturing. The industrial revolution introduced the idea of using machines to increase production while reducing costs.

In recent years, there has been a rapid increase in the use of robots in manufacturing [2]. Robots have changed manufacturing in a myriad of positive ways. Robotic automation introduces the idea of using machines or robots to increase production while lowering costs, reducing worker fatigue, and increasing precision, accuracy, repeatability, and safety.

© The Author(s), under exclusive license to Springer Nature Switzerland AG 2023
M. N. O. Sadiku et al., *Emerging Technologies in Manufacturing*,
https://doi.org/10.1007/978-3-031-23156-8_3

This chapter provides an introduction to the uses of robotic automation in manufacturing. It begins by describing what robots are. It gives an overview on automation. It describes different types of robots and how they are used in manufacturing. It highlights the benefits and challenges of using robots in manufacturing. It covers global adoption of robotic automation. The last section concludes with comments.

3.2   What Is a Robot?

The word "robot" was coined by Czech writer Karel Čapek in his play in 1920. The Russian writer Isaac Asimov coined the term "robotics" in 1942 and came up with three rules to guide the behavior of robots [3]:

1. Robots must never harm human beings.
2. Robots must follow instructions from humans without violating rule 1.
3. Robots must protect themselves without violating the other rules.

The first robots had no external sensing and were used for simple tasks as pick and place. They replaced humans in monotonous, repetitive tasks, and they were deployed in dangerous work environments. Robotics has since advanced and taken many forms including fixed robots, collaborative robots, mobile robots, industrial robots, medical robots, police robots, military robots, officer robots, service robots, space robots, social robots, personal robots, and rehabilitation robots [4, 5]. Robots are becoming increasingly prevalent in almost every industry, from healthcare to manufacturing.

Although there are many types of robots designed for different environments and for different purposes/applications, they all share four (4) basic similarities [6]: (1) All robots have some form of mechanical construction designed to achieve a particular task. (2) They have electrical components which power and control the machinery. (3) All robots must be able to sense their surroundings; a robot may have light sensors (eyes), touch and pressure sensors (hands), chemical sensors (nose), hearing and sonar sensors (ears), etc. (4) All robots contain some level of computer programming code. Programs are the core essence of a robot since they provide intelligence. There are three different types of robotic programs: remote control, artificial intelligence, and hybrid. Some robots are programmed to invariably and devotedly carry out specific actions over and over again (repetitive actions) without variation and with a high degree of accuracy. Some advantages and disadvantages of robots are shown in Fig. 3.1 [7]. Robots are used in many areas such as food manufacturing, electronic manufacturing, business, agriculture, healthcare, automotive, transportation, and law enforcement.

Robotics is a branch of engineering and computer science that involves the conception, design, manufacture, and operation of robots. It is an interdisciplinary advanced technology field, embracing mechanical engineering, electrical engineering, computer science, biotechnology, nanotechnology, information technology, and robot technology. It may be regarded as one of the technologies used to design

Fig. 3.1 Some advantages and disadvantages of robots [7]

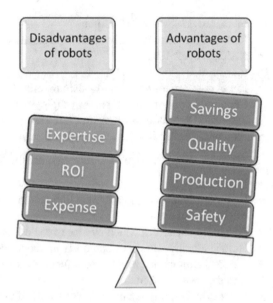

and operate robots. The goal of robotics is to create intelligent machines (called robots) that behave and think like humans.

3.3 Overview on Automation

Automation refers to a wide range of technologies that reduce human intervention in processes. It has been achieved by various means including mechanical, hydraulic, pneumatic, electrical, electronic devices, and computers. There are different types of automation including artificial neural network, human machine interface, robotic process automation, supervisory control and data acquisition (SCADA), programmable logic controller (PLC), and robotics. Automation can also be classified as: (1) fixed automation, (2) programmable automation, and (3) flexible automation.

Robots exhibit varying degrees of autonomy. Autonomy means to be independent and able to govern oneself. It is different from automation, which performs a sequence of highly structured preprogrammed tasks. Industrial autonomy is where plant assets and operations have learning and adaptive capabilities that allow responses with minimal human interaction. Some companies are transitioning from industrial automation to industrial autonomy.

The main advantages of automation are the following [8]:

- Increased throughput or productivity
- Improved product quality
- Increased predictability
- Improved robustness (consistency) of processes or product

- Increased consistency of output
- Reduced direct human labor costs and expenses
- Reduced cycle time
- Increased accuracy
- Relieving humans of monotonously repetitive work
- Required work in development, deployment, maintenance, and operation of automated processes—often structured as "jobs"
- Increased human freedom to do other things

The main disadvantages of automation are the following [8]:

- High initial cost.
- Faster production without human intervention which can mean faster unchecked production of defects.
- Scaled-up capacities which can mean scaled-up problems when systems fail.
- Human adaptiveness is often poorly understood by automation initiators.
- Workers anticipating employment income may be seriously disrupted by companies deploying automation.
- Current technology is unable to automate all the desired tasks.
- Many operations using automation have large amounts of invested capital and produce high volumes of product, making malfunctions extremely costly and potentially hazardous.
- As a process becomes increasingly automated, there is less and less labor to be saved or quality improvement to be gained.
- As more and more processes become automated, there are fewer remaining non-automated processes.

To jump directly to autonomous operations is hard to achieve. The following levels of automation must be taken [9].

Level 0–1 Manual/Semi-automated: A facility is minimally instrumented and automated to improve productivity. Many operations are performed manually with paper-based instructions and record keeping.

Level 2 Automated: The automation system conducts majority of production processes but requires human oversight and intervention.

Level 3 Semi-autonomous: It is characterized by a mixture of autonomous components and automated assets with human orchestration. Companies at this level deploy a range of selective autonomous components or applications orchestrated by humans.

Level 4 Autonomous Orchestration: Most assets operate autonomously and are synchronized to optimize production, safety, and maintenance. There is still a need for humans to perform many tasks.

Level 5 Autonomous Operations: A highly idealized state where facilities operate autonomously and require no human interaction.

Although perfect automation has never been realized, it has caused alterations in the patterns of employment.

3.4   Robotics in Manufacturing

Productivity, cost-efficiency, reliability, reduced dependency, and sustainability are some of the key factors driving the adoption of robotic automation in today's manufacturing environment. Here are some types of robotic technology that have changed and will keep changing the manufacturing industry [10]:

1. *Collaborative Robots*: These robots are designed to work collaboratively with people in manufacturing environments.
2. *Autonomous Mobile Robots*: These robots can get around by themselves in manufactory environment.
3. *Industrial Robots*: These robots have strong arms that can handle massive payloads in factories.
4. *Robots with Machine Vision*: These robots use machine vision to help quality control workers and inspect parts.
5. *Robotic Blacksmithing*: This manufacturing method uses a combination of special tools, robotic arms, and sensors to reshape materials.
6. *Mobile Robots*: Robots are either fixed or mobile. Mobile robots are designed for changing environments. They are not bolted to the floor as fixed robots.
7. *Articulated Robots*: Articulated robots have rotary joints that allow for a full range of motion. They are industrial robots used in manufacturing for packing, painting, welding, and material handling.

These robots are evolving. They are growing in size, capability, and speed. The rapidly growing number of robots has been used for professional and personal service applications.

Robots are used in manufacturing for the following reasons [11]:

1. To create and enhance efficiencies all the way from raw material handling to finished product packing.
2. They can be programmed to operate 24/7 in lights-out situations for continuous production.
3. Robotic equipment is highly flexible and can be customized to perform even complex functions.
4. Manufacturers increasingly need to use robotic automation to boost productivity and stay competitive.
5. Robotic automation can be highly cost-effective for nearly every size of company.
6. Any repetitive task is a candidate for robotic manufacturing. Robots protect workers from repetitive, mundane, and dangerous tasks.
7. Robots handle tiny parts too small for human eyes and never make mistakes.
8. Robots free up manpower to let companies maximize workers' skills in other areas of the business. They create more desirable jobs, such as engineering, programming, management, and maintenance.
9. Robotic automation allows domestic companies to be price-competitive with offshore companies.

10. Robots achieve ROI quickly, often within 2 years, offsetting their upfront cost.

Robots are used in manufacturing to perform different functions. The most common areas where robots are performing their jobs in the manufacturing process include [12]:

1. *Material Handling*: Robots are being used to handle materials that require dangerous product that could risk contamination if in contact with humans.
2. *Welding:* The process of joining metal pieces is a dangerous and risky job that requires exact precision. Robots are becoming a popular choice for welding jobs. Welding robots are shown in Fig. 3.2 [13].
3. *Assembly:* Having to assemble product parts is a long, repetitive job. By replacing such a system with a robot significantly reduces error.
4. *Dispensing:* For processes which require glue, paint, or sprays, dispensing robots are placed at a strategic point near the path of the product.
5. *Processing:* There are certain products that have to undergo a specific type of processing, such as carving, polishing, or sawing, before being released. This task is done by robots with varying degrees of autonomy.
6. *Packaging*: Robots are used in packaging in a wide range of industries because they offer many product protection and safety, traceability, logistics, and supply chain benefits. A typical packing robot is shown in Fig. 3.3 [14].

Fig. 3.2 Welding robots [13]

Fig. 3.3 A packaging robot [14]

3.5 Manufacturing Applications

Robotic automation can be applied to many different areas in manufacturing. The most common ways robotic automation is used in manufacturing include the following:

- *Automotive Industry:* This is the largest user of robots in advanced nations around the world. In particular, it is the largest customer of industrial robots. Robots are used in almost every aspect of automotive manufacturing. Robots are more efficient, accurate, flexible, and dependable on production lines. Robotic automation has allowed the automotive industry to remain one of the most automated supply chains globally. Different ways that robots are helping automotive manufacturers improve their automation processes include robotic vision, spot and arc welding, assembly, painting, sealing and coating, machine tending and part transfer, materials removal, and internal logistics [15]. A typical automotive manufacturing system is shown in Fig. 3.4 [16].
- *Electronics Manufacturing:* Electronics manufacturing is increasingly becoming complex as the size of components and circuits continue to shrink. Robotic automation has great potential in the manufacturing of today's sophisticated electronic devices and products. It applies to almost all the stages in the electronics production cycle. It delivers a wide range of cost, quality, flexibility, risk-reduction, and safety benefits. Typical functions of robotic automation include material and component handling, assembly lines, etching, inspections, soldering, visual and physical testing, component fabrication, pick and place operations, assembling miniature components on PCBs, applying adhesives,

Fig. 3.4 A typical automotive manufacturing [16]

inspections, and packaging. The robots can reduce the labor costs significantly by cutting on the number of employees while increasing the production times and reducing errors and wastage. Robots with arm-mounted cameras can visually inspect electronics assemblies. Fig. 3.5 shows how a robot is used in electronic manufacturing [17].

- *Lights-Out Manufacturing:* This is a production system with no human workers; machines handle the production process from beginning to end to eliminate labor costs. The "lights out" manufacturing concept is so called because robots can work without lights, do not need HVAC, coffee breaks, or days off and, indeed, do not need other conditions that human workers require. Lights-out manufacturing allows robots to work without any interference. It became popular in 1982 when General Motors replaced risk-averse bureaucracy with automation and robots. The expansion of lights-out manufacturing requires reliability of equipment, preventive maintenance, and commitment from the staff. Companies that practice this manufacturing style can experience better energy efficiency [7].
- *Automated Production Lines:* Automating tedious and difficult tasks with robot revolutionized the manufacturing sector. Automation has been achieved through a combination of various means of technology and system integration including mechanical, hydraulic, pneumatic, electrical, electronic, and computers. Today, extensive automation is practiced in almost every type of manufacturing. Automated production line consists of a series of workstations connected by a transfer system to move parts between the stations. This is an example of fixed automation, since these lines are typically set up for long production runs. The

Fig. 3.5 Robot is used in electronic manufacturing [17]

various operations and other activities taking place on an automated transfer line must all be sequenced and coordinated properly for the line to operate efficiently. Automated production lines are used in many industries, especially automotive industry [18].

• *Robotic Processing Automation:* The manufacturing industry is leading the way as it increasingly adopts robotic process automation (RPA). RPA allows manufacturers to automate certain types of work processes to reduce the time spent on costly manual tasks. It is a critical innovation within Industry 4.0. It can automate a host of repetitive, rules-based processes, minimizing the amount of time spent on manual tasks, improving productivity, driving innovation, and lowering costs. Unfortunately, switching RPA platforms has proven to be both expensive and difficult to execute to date due to numerous other challenges. As RPA technology improves, manufacturers' RPA portfolios are likely to see a formidable rise in growth. Companies focus on getting their RPA functioning with increased stability and fewer errors [19]. Figure 3.6 illustrates RPA in the manufacturing industry [20]. Some of the benefits of RPA in the manufacturing industry are illustrated in Fig. 3.7 [21] and explained as follows [21, 22]:

• *Accurate:* RPA software is generally less prone to errors and function with high precision.
• *Consistent*: This software can produce error-free, consistent results.

Fig. 3.6 RPA in the manufacturing industry. (Modified [20])

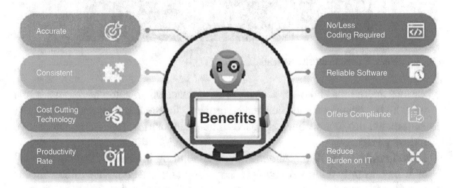

Fig. 3.7 Benefits of RPA [21]

- *Cost Cutting Technology:* RPA definitely reduces the manual workforce and hence reduces the cost used to perform any task.
- *Productivity Rate*: The execution time to perform any task is much faster when compared to the manual approach.
- *Reduce Burden on IT*: RPA software does not disturb the underlying legacy systems.
- *Offers Compliance:* RPA follows rules to prove audit-free trial so that you do not have to worry about the security of your automation.
- *Reliable Software:* This software is reliable, as robots can work 24/7 without a break.
- *No/Less Coding Required*: RPA software require a minimal level of programming knowledge.

Other manufacturing sectors using robotic automation include food manufacturing, forging, forming, shaping and blanking manufacturing industry, reshoring manufacturing, and robot welding.

3.6 Benefits

Robots are evolving in ways many business professionals and production managers across the globe could only have dreamed of. Automation and robotics continue to influence our lives in many ways. They make manufacturing easier, more productive, and more efficient. The key objective of robotic automation is to improve worker safety, reduce costs, improve quality, and increase flexibility. Robotic autonomous operation is a means to achieve smart manufacturing goals. The advantages of robotics include the efficient execution of heavy-duty jobs with precision and repeatability. A robot increases speed for manufacturing processes by operating 24/7, thereby increasing production while minimizing downtime. Other benefits of robotic automation include [19]:

1. *Better Quality and Consistency*: Robots can provide better production quality and more precise and reliable processes. They can increase productivity, efficiency, and safety while delivering risk-reduction during process operation.
2. *Maximum Productivity:* Higher output and increased productivity have been two of the fundamental economic advantages commonly attributed to automation. The productivity of a process, which is greatly enhanced by robotic operations, is traditionally defined as the ratio of output units to the units of labor input.
3. *Replacing Repetitive Tasks:* A repetitive job can be better handled by a robot than a human worker simply due to lack of errors. Using robots for repetitive tasks means fewer risks of injury for workers.
4. *Automation:* By automating some manufacturing processes, labor is eliminated. Automation increases manufacturing productivity, economic competition, and education. It is also associated with improved quality, fewer errors, reduced safety instances, faster manufacturing production, and cheaper labor.
5. *Reduced Direct Labor Costs:* Replacing some workers with robots frees up workers so their skills and expertise can be used somewhere else. Reduction in labor increases profit, which is always a business goal imperative.
6. *Addressing Challenges:* Robots are helping manufacturers address many of the key challenges they face, including tight labor pools, global market competitiveness, and safety.
7. *Minimize Cost:* Robotic automation is becoming increasingly flexible and intelligent; the realization of which is minimizing cost per unit.
8. *Increased Efficiency:* In manufacturing, robots create efficiencies all the way from raw material handling to finished packaged product.
9. *Programmable:* Robots can be taught through programming to perform certain tasks, following a sequence of steps. They are getting smarter.

10. *Longer work hours:* Industrial robots work at a faster speed without fatigue or the need to take breaks or pauses.

3.7 Challenges

Like all things, robots come with pros and cons. There are still challenges of robotic automation that must be overcome. These challenges include [23]:

1. *High initial investment:* Robots typically require a large upfront investment. Manufacturing companies must be careful not to overcomplicate or overspend on their investment on robotic automation.
2. *Expertise can be scarce:* Deploying robot requires training and developing expertise. Industrial robots need sophisticated operation, maintenance, and programming. The number of people with these skills is currently limited.
3. *Ongoing costs:* While industrial robots may reduce some manufacturing labor costs, they do come with their own ongoing expenses, such as maintenance.
4. *Competition:* Many manufacturers have been compelled to send jobs offshore to lower-wage manufacturing job jurisdictions, because they could not compete with low-cost foreign labor.
5. *Labor:* Robotic automation involves a replacement of human labor by an automated system. Workers have indeed lost jobs through automation.
6. *Stress:* A worker whose job has been replaced by robots goes through a period of emotional stress. The worker may need to relocate.
7. *Limitations of Robots:* Although robots remain an attractive alternative over human labor, there are still tasks that robots cannot perform. Robots depend upon their enabling systems, such as vision systems, grippers, conveyors, motion-sensor systems, and PLCs to complete tasks.
8. *Robotic Integration:* The benefits of integrating robotics into an organization's production or manufacturing lines are endless. However, integrating robotics into a manufacturing system is a complex task. It requires judicious investment, predicated on calculating the risk and return tradeoffs and the implicit and explicit ROI and embracing changes based on both certain and uncertain limitations.
9. *Societal Acceptance:* The problem with robots is not just an issue of appropriate technology availability, choice, and insertion but also one of heavy dependence on societal acceptance, safety, and reliability as well as government regulations and industrial standardizations. Robots can be dangerous, and absolute safety is not guaranteed with their use. "Safe robots" should be built that do not harm workers in case of an accident or collision.
10. *Ethical Concerns:* The legal and ethical barriers are high. Advances in robotic technology have social, legal, and ethical ramifications. Are robots primarily introduced to solve problems in manufacturing or replace humans in order to save money? Robots lack the capacity of moral reasoning. Ethical issues such as personal moral responsibility, privacy, and accountability are emerging con-

cerning the use of robots. This suggests the need for developing robot ethics and examining the laws and regulations governing its use [24, 25].

3.8   Global Adoption of Robotic Automation

Industrial globalization of trade has increased and changed the manufacturing sector worldwide. The use of robots has expanded globally. Germany and Italy are ahead of the United States in terms of adoption of robot technology in production. The International Federation of Robotics (IFR) has robot immersion by industry and sector data for 13 industries within manufacturing and for 6 broad sectors outside of manufacturing. According to IFR, there were about 2,439,543 operational industrial robots by the end of 2017. We now consider how robotic automation is deployed in many countries.

- *United States: As* an area of critical importance to the United States and its economy, the manufacturing sector represents the largest of the United States' private industry sectors. Robotics has made North American manufacturers globally competitive. The US auto industry employed 136 robots per thousand workers, while all other manufacturing industries in the United States employed only 8.6 robots per thousand workers. Interest in robotics increased in the late 1970s, and many US companies entered the robotic system deployment wagon including General Electric and General Motors. Unemployment is becoming a serious social problem in the United States due to the exponential growth rate of automation and technological advances.
- *China:* This is the largest industrial robot market, with 154,032 units sold in 2018. In China, which is a manufacturing hub due to cheap labor, most factories have replaced half of their workforce with robots.
- *Japan:* FANUC, a leading robot manufacturer in Oshino, Japan, has a 22-factory complex where they use the *lights-out manufacturing* concept to build their products. Industrial robots are supervised by a staff of only four workers per shift. In Japan, trials have demonstrated that robots can reduce the time required to harvest strawberries by up to 40 percent.
- *Canada:* The Canadian advanced manufacturing sector includes the fields of robotics, 3D printing, and ICT technologies. Canada facilitates the seamless integration of Industry 4.0 solutions into manufacturing operations [26].

3.9   Conclusion

Robots are all around us and their uses are increasing every day. They are taking over the world. They are designed to perform a wide variety of programmed tasks. Manufacturing and robotics are an inextricably tied, indissolubly bound, and naturally collaborating duo. They hold great promise and are a key to projected rapid, sustained growth in industrial productivity. Robots are rapidly changing the face of manufacturing. They may help manufacturers increase precision, repeatability, and productivity. Today, manufacturing robots are more affordable than ever before.

Every manufacturer can benefit from employing robots.

The manufacturing industry is constantly changing. The future of manufacturing is becoming brighter, yet sophisticated, due to the various challenges manufacturers face and several mind-blowing devices and technologies in the horizon that would be available for insertion.

In the future, as demand for goods continues to grow, these robots may become more common on manufacturing floors. Their cost as well as the cost of adoption will fall. Future generations of robots are likely to offer higher levels of precision. Without a doubt, robots will play a pivotal role in the future of manufacturing. More information about robotic automation can be found in the books in [27–35], on a website on robots (robots.ieee.org), and from the following journals devoted to robot-related issues:

- *Advanced Robotics*
- *Journal of Robotic Systems*
- *Journal of Robotics*
- *Journal of Robotic Surgery*
- *Journal of Intelligent & Robotic Systems*
- *Intelligent Service Robotics*
- *IEEE Journal on Robotics and Automation*
- *IEEE Robotics & Automation Magazine*
- *IEEE Transactions on Robotics*
- *International Journal of Medical Robotics and Computer Assisted Surgery*
- *International Journal of Robotics Research*
- *International Journal of Social Robotics*
- *International Journal of Humanoid Robotics*
- *Robotics and Autonomous Systems*

References

1. M.N.O. Sadiku, U.C. Chukwu, A. Ajayi-Majebi, S.M. Musa, Robotic automation in manufacturing. J. Sci. Eng. Res. **8**(4), 140–149 (2021)
2. A. Bharadwaj and M.A. Dvorkin, The rise of automation: How robots may impact the U.S. labor market (2019), https://www.stlouisfed.org/publications/regional-economist/second-quarter-2019/rise-automation-robots
3. I. Asimov, *Robots and Empire* (Grafton Books, London, 1985)
4. R.D. Davenport, Chapter 3: Robotics, in *Smart Technology for Aging, Disability, and Independence*, ed. by W.C. Mann, (Wiley, New York, 2005), pp. 67–109
5. M.N.O. Sadiku, S. Alam, S.M. Musa, Intelligent robotics and applications. Int. J. Trends Res. Dev. **5**(1), 101–103 (2018)
6. Robotics, *Wikipedia*, the free encyclopedia (n.d.), https://en.wikipedia.org/wiki/Robotics
7. L. Calderone, Robots in manufacturing applications (2016), https://www.manufacturingtomorrow.com/article/2016/07/robots-in-manufacturing-applications/8333

8. Automation, *Wikipedia*, the free encyclopedia (n.d.), https://en.wikipedia.org/wiki/Automation
9. IA2IA (n.d.), https://www.yokogawa.com/us/solutions/solutions/ia2ia/
10. 5 Robots changing manufacturing forever (2020), https://www.robotshop.com/community/blog/show/5-robots-changing-manufacturing-forever
11. Robotics in manufacturing (n.d.), https://www.acicta.com/why-robotic-automation/robotics-manufacturing/
12. How are robots used in the manufacturing industry? (n.d.), https://www.maderelectricinc.com/blog/how-are-robots-used-in-the-manufacturing-industry
13. Robotic industry news (n.d.), https://www.robots.com/articles/arc-welding-101
14. An overview of packaging robots benefits and capabilities (n.d.), https://www.genesis-systems.com/blog/an-overview-of-packaging-robots-benefits-and-capabilities
15. 7 Key robot applications in automotive manufacturing (n.d.), https://www.roboticsbusinessreview.com/manufacturing/7-key-robot-applications-in-automotive-manufacturing/
16. Robotics Industry Insights – The Art Of (n.d.), https://www.wrerw.xyz/ProductDetail.aspx?iid=106601332&pr=45.99
17. A. Kingatua, Robots and automation in electronics manufacturing (2020), https://medium.com/supplyframe-hardware/robots-and-automation-in-electronics-manufacturing-a77f177585eb
18. Manufacturing applications of automation and robotics (n.d.), https://www.britannica.com/technology/automation/Manufacturing-applications-of-automation-and-robotics
19. T. Higgins, How robotic processing automation will change in 2021 (2021), https://www.manufacturing.net/automation/blog/21319201/how-robotic-processing-automation-will-change-in-2021
20. Y. Devarajan, A study of robotic process automation use cases today for tomorrow's business. Int. J. Comput. Tech. **5**(6) (2018)
21. S. Kappagantula, Robotic process automation – All you need to know about RPA (2020), https://www.edureka.co/blog/robotic-process-automation/
22. The benefits of robotic process automation (n.d.), https://www.laserfiche.com/ecmblog/what-is-robotic-process-automation-rpa/
23. M. Stevens, Pros and cons of using industrial robots in your manufacturing operation (2019), https://www.wipfli.com/insights/blogs/manufacturing-tomorrow-blog/pros-and-cons-of-industrial-robots-in-manufacturing
24. K. Kernaghan, The rights and wrongs of robotics: Ethics and robots in public organizations. Can. Public Adm. **57**(4), 485–506 (2014)
25. B. CarstenStahl, M. Coeckelbergh, Ethics of healthcare robotics: Towards responsible research and innovation. Robot. Auton. Syst. **86**, 152–161 (2016)
26. Advanced manufacturing (n.d.), Unknown Source
27. T.R. Kurfess, *Robotics and Automation Handbook* (CRC Press, Boca Raton, 2018)
28. D. Zhang, Z. Gao (eds.), *Recent Developments in Manufacturing Robotic Systems and Automation* (Bentham Books, 2013)
29. M.A. Boboulos, *Automation and Robotics* (Ventus Publishing, 2010)
30. T. Bock, T. Linner, *Robotic Industrialization* (Cambridge University Press, 2015)
31. M.K. Habib, *Advanced Robotics and Intelligent Automation in Manufacturing* (IGI Global Publisher, 2020)
32. A.G. Kravets (ed.), *Robotics: Industry 4.0 Issues & New Intelligent Control Paradigms* (Springer, 2020)
33. P. Mckinnon, *Robotics: Everything You Need to Know About Robotics from Beginner to Expert* (CreateSpace Independent Publishing, 2016)
34. B. Siciliano et al., *Robotics: Modelling, Planning and Control (Advanced Textbooks in Control and Signal Processing)* (Springer, 2009)
35. A. M. Tripathi, *Learning Robotic Process Automation: Create Software Robots And Automate Business Processes With the Leading RPA Tool – Uipath: Create Software Robots ... With the Leading RPA Tool.* Packt Publishing, 2018

Chapter 4
Smart Manufacturing

No country is ever successful in the long term... without a really strong and vibrant manufacturing base.

–Alan Mulally.

4.1 Introduction

Traditionally, manufacturing has been regarded as the process of converting raw materials into finished product. It is essential for realizing all future products and is an indispensable element of the innovation chain. The manufacturing industry includes many sectors such as chemical industry, metal machinery industry, and electronics industry. Manufacturing has gone through transformations and become more automated, computerized, connected, and sophisticated.

Traditional manufacturing companies currently face several challenges such as rapid technological changes, inventory problems, shortened innovation span, short product life cycles, volatile demands, low prices, highly customized products, and the ability to compete in the global markets. The increasing impact on the manufacturing industry of Information Communication Technologies (ICT), encompassing the internet, computers, hardware and software, wireless networks, cell phones, middleware, modems, routers, repeaters, video-conferencing, social networking, and many other media applications and services. Digital technologies are leading to the development of new methods and tools that support the adoption of new production and product development technologies. The manufacturing system is moving toward a more reliable and intelligent direction and end. The future of manufacturing is geared toward producing customer-specific products ensuring short product life cycles, quick delivery times, zero defect production, and resource-efficient manufacturing [1]. The features can be achieved by smart manufacturing, which is also known as intelligent manufacturing. Due to environmental and resource limitation, smart manufacturing seems to be the only way to improve the level of productivity, quality, reliability, safety, and serviceability of manufacturing systems.

Select and innovative technologies are transforming the manufacturing industry. In particular, smart manufacturing (SM) has gained significant impetus as a breakthrough technological development that can transform the landscape of manufacturing today and tomorrow. It aims to take advantage of advanced information and manufacturing technologies to enable flexibility in physical processes to address a dynamic and global market. It is becoming the focus of global manufacturing transformation and upgrading [2].

This chapter provides an introduction to smart manufacturing variously referred to as the smart factory. It begins by covering traditional manufacturing. Then it introduces the concept and characteristics of smart manufacturing. It discusses the goals of smart manufacturing and its enabling technologies. It provides some applications of SM, devotes a section to smart factory, presents some benefits and challenges on SM, and covers the global adoption of SM. The last section concludes with comments.

4.2   Overview of Traditional Manufacturing

Manufacturing started from a small-scale production line of crafts in 1800s. It has today evolved into large-scale mass production. The 2000s is the era when computerized and personalized manufacturing systems came into existence for mass production lines [3]. Manufacturing is typically a series of processes comprising selection of raw materials, production of objects, assembling of parts, inspection, and dispatching [4]. It may include foundry, forging, joining, welding, grinding, milling, turning cutting, boring, rolling, forming, abrading, polishing, electropolishing, deburring, masking, ion-implantation, heat treatment, painting, etc. It involves using shop floor resources to meet the delivery date, cost, quality, and optimal economic goals in limited resources condition. It often involves mass production and heavy energy consumption resources such as coal or electricity.

The traditional or current manufacturing processing techniques consume lots of energy, mainly for production and utility. They also produce a lot of pollution and add to the deterioration of the global environment. Because of this, manufacturers are gradually transforming their manufacturing systems from traditional mass production to flexible lean systems. With rapid changes in technology, manufacturing itself is constantly transforming and evolving [5]. It now takes a proactive role in the development of cleaner manufacturing processes. In order to minimize the environmental damage due to manufacturing, there is a need for new manufacturing process.

In modern times, two types of manufacturing systems have emerged emphasizing waste minimization. They are "lean" manufacturing systems and "green" manufacturing systems that both reduce waste. Lean manufacturing seeks to eliminate all types of wastes generated within a production system. In lean manufacturing, there are eight categories of waste that should be monitored [6]: (1) overproduction, (2) waiting, (3) inventory, (4) transportation, (5) over-processing, (6) motion, (7) defects, and (8) workforce. Green manufacturing (GM), also known as environmentally conscious manufacturing, refers to modern manufacturing that makes products without pollution. It addresses a wide range of environmental and sustainability issues including resource selection, transportation, manufacturing process, and pollution. It is an effective way to protect resources and the environment and conserve the resources for future generation.

4.3   Concept of Smart Manufacturing

Recent advances in the field of technology have led to the emergence of innovative solutions known as smart technologies. A technology is considered smart if it performs a task that an intelligent person can do. A smart or intelligent technology is a self-operative and corrective system that requires little or no human intervention. Smart technologies can be understood as a generalization of the concept of smart structures and the use of digital and communications technologies. They have given users new, powerful tools to work with. Today, we are in an era where everything is expected to be smart. Common examples include smart cities, smart factory, smart agriculture, smart farming, smart healthcare, smart university, smart medication, smart water, smart food, smart materials, smart devices, smart phones, smart grid, smart energy, smart homes, smart buildings, smart metering, smart appliances, smart equipment, smart heating controls, smart lighting systems, smart watch, smart economy, smart environment, smart grids, smart transportation, smart mobility, smart manufacturing, smart living, smart environment, smart people, etc. These technologies will ensure equity and fairness and lead to the realization of a better society and quality of life [7].

Today, manufacturers are facing the challenges of shortened product lifecycle, increasing customer-oriented value, global competition, and sustainable development. Smart manufacturing (SM) is envisioned as a good approach to tackling these challenges. It may be regarded as an integration of multiple users, multiple systems, and multiple technologies of information processing. It is expected to have the abilities to assist in decision-making, to adapt to new situations, and correct fabrication problems in a dynamic manufacturing environment.

Smart manufacturing (SM) is a term coined by several agencies in the United States and is increasingly being used globally. Other initiatives such as Industry 4.0, 5G, cyber-physical production systems, smart factory, intelligent manufacturing, and advanced manufacturing are frequently used synonymously with smart manufacturing. SM is the integration of new-generation information communication technology (ICT) and the manufacturing industry. The core of achieving integration refers to the Cyber-Physical Systems (CPS) capable of connecting the real physical world and virtual network world. These and other technologies enable a manufacturing system to become "smart." Thus, smart manufacturing is the right data in the right form, the right people with the right knowledge, the right technology and the right operations, whenever and wherever needed throughout the manufacturing enterprise [8]. It is a manufacturing type in which all information is available when it is needed, where it is needed, and in the form it is most useful. Figure 4.1 depicts typical smart/industrial manufacturing [9].

Fig. 4.1 Typical smart/industrial manufacturing [9]

Smart manufacturing will transform traditional manufacturing from cost operations into added-values operations and increase competitiveness. It is a value-creation process from design to production, logistics, and service. Through SM, manufacturing is shifting from knowledge-based intelligent manufacturing to data-driven and knowledge-enabled smart manufacturing [10]. In SM systems, data analytics plays an important role in turning data into valuable insights to assist decision-making.

Interest in smart manufacturing has continued to increase under different terms: smart manufacturing in the United States, Industry 4.0 in Germany, and Smart Factory in Korea. This indicates several initiatives of the same concept in many locations across the globe [11]. As illustrated in Fig. 4.2, smart manufacturing represents the fourth industrial revolution, which has arrived after the invention of the steam engine, mass production, and industrial automation [12]. The fourth industrial revolution is transforming global manufacturing. It is based on the application of Internet of things and other concepts into manufacturing companies. The resulting smart factories are able to fulfill dynamic customer demands while integrating human ingenuity and automation.

4.4   Characteristics of Smart Manufacturing

The term "smart manufacturing" (SM) originated in the United States and is increasingly used globally. Smart manufacturing, which is the fourth revolution in the manufacturing, has its own identity captured in the following six pillars [13], shown in Fig. 4.3: (1) manufacturing technology and processes, (2) materials, (3) data, (4) predictive engineering, (5) sustainability, and (6) resource sharing and networking. Three

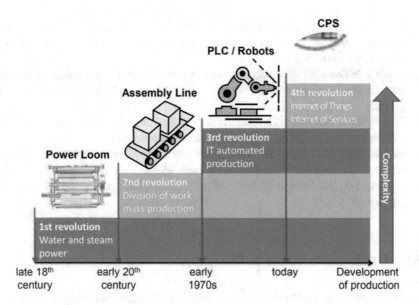

Fig. 4.2 Four industrial revolutions [12]

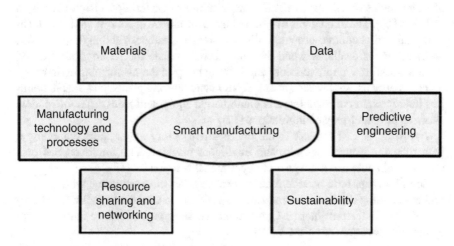

Fig. 4.3 *Six pillars of smart manufacturing* [13]

more pillars can be added [14, 15]: (7) improved efficiency, (8) risk reduction, (9) and agile decision-making. Smart manufacturing (SM) systems include data collection, data management, and data analysis, which provide real-time actionable information required to build system intelligence into manufacturing operation.

Smart manufacturing often involves making available the right data in the right form, for the right people with the right knowledge, the right technology and the right operations. Thus, smart manufacturing system covers four areas [14, 15]:

1. *Smart Parts:* The sensor technique is applied to the components so that they can have self-sensing capability of sensing temperature and vibration to improve reliability and service life.
2. *Smart Machine:* This collects information on the operating status of various parts of the machine, conduct big data analysis, detect abnormalities as early as possible.
3. *Smart Production Line:* The machines on the production line can communicate and support each other to increase the production efficiency.
4. *Smart Factory:* This is linked with consumer needs and provides customized services. The Internet is used to link the management systems between factories to control the production schedule of products at linked or cooperating machining centers.

4.5   Goals of Smart Manufacturing

Modern manufacturing industry seeks to improve competitiveness through the convergence with cutting-edge ICT technologies. Smart manufacturing is the collection of technologies that support effective and accurate engineering decision-making in real time [16]. The main goals of smart manufacturing are as follows: to foster technological and economic growth, build a customer-awareness and agile platforms, make resources available when they are needed, create the required skilled and trained workforce, improve safety and sustainability in the manufacturing industry, reduce energy use and waste streams from manufacturing plants, and enable seamless interoperation of manufacturing automation equipment from different vendors allowing plug-and-play configurations [17].

Applying the SM concept to the future smart factory may require equipping it with robots, advanced sensors, and intelligent machine. The improved SM processes will handle and manage more operational complexity [18].

Several groups have been formed to advance smart manufacturing, with the most prominent being the Smart Manufacturing Leadership Coalition (SMLC), Industry 4.0, and the Industrial Internet Consortium. These groups comprise industry, academic, and government partners [19].

4.6   Enabling Technologies

The implementation of the smart manufacturing/factory is enabled by several technologies such as artificial intelligence, industrial robots, embedded systems, RFID, sensor networks, wireless connectivity, machine learning, big data, cloud computing, cyber physical system (CPS), blockchain, human-machine interaction (HMI), augmented reality (AR), 3D printing, digital twin, and the Internet of things (IoT) or industrial Internet of things (IIoT). Thus, smart manufacturing enabling technologies include [20] the following:

1. *Cyber-physical system* (CPS) is a key technology for smart manufacturing. CPS consists of physical entities (such as machines, vehicles, and work pieces), which are equipped with technologies such as RFIDs, sensors, microprocessors, telematics, or embedded systems. The Internet of things (IoT) is essentially the connection of machines, devices, and "things" wirelessly. IoT allows devices to be monitored and controlled remotely over the communication network. It has shown great potential in device communication, connections, and data collection, thereby laying a solid foundation for cyber–physical integration.

2. *Sensors:* Ubiquitous sensing technologies include radio frequency identification (RFID), auto ID, virtual reality, GPS, and Wi-Fi. A regular sensor can be converted into a ubiquitous sensor by connecting a networking module to it. The sensors (temperature, vibration, force, etc.) are used to monitor manufacturing processes.

3. *RFID*: Manufacturing resources like machines, materials, and personnel are equipped with RFID devices and they become smart manufacturing objects. RFID technology enables real-time traceability, visibility, and interoperability in improving the performance of shop-floor planning, execution, and control of manufacturing systems.

4. *Robots*: Industrial mobile robots are used in handling and transporting materials. This is appropriate for smart factory due to their flexibility and ability to communicate. Robots get the data from sensors and change their action accordingly. Robots with AI capability make it possible to involve the perception-based decision-making which otherwise was not possible by rule-based algorithms in robots.

5. *IoT*: There is huge potential for IoT and IIoT in smart manufacturing. IoT allows any individual object or device to be connected to the Internet. It is a link between objects in the real world with the virtual world, thereby enabling anytime, anywhere connectivity for anything. Major forces driving IoT in smart manufacturing are the growing need for centralized monitoring and predictive maintenance of manufacturing infrastructure. While the IoT affects transportation, healthcare, or smart homes, the Industrial Internet of Things (IIoT) refers in particular to industrial environments. IIoT refers to the application of IoT across several industries such as manufacturing, logistics, oil and gas, transportation, energy/ utilities, chemical, aviation, and other industrial sectors. This killer application is poised to revolutionize the world.

6. *3D printing:* Additive manufacturing (AM), also known as 3D printing, is one of the most revolutionary technologies in the history of manufacturing. It is basically used in rapid prototyping. This layer-by-layer production technology is rapidly finding an increased number of applications in manufacturing.

Some of these enabling technologies are shown in Fig. 4.4 [21].

Fig. 4.4 Some enabling technologies *for smart manufacturing* [21]

4.7   Applications

Manufacturing has evolved and become more automated, computerized, connected, and complex. It is an emerging form of production integrating manufacturing assets of today and tomorrow. Some typical applications of smart manufacturing include the following:

- *Chemical Industry*: The chemical industry can apply smart manufacturing for chemicals production optimized to have targeted final-use properties requested by the customer. For the chemical process industry, smart manufacturing should not only maximize economic competitiveness but also significantly reduce accident occurrences, while conversely enhancing safety compliance records and experiences.
- *Semiconductor Manufacturing*: The semiconductor manufacturing technologies are the prime mover of consumer electronics market. It is part of a larger manufacturing discipline often referred to as microelectronics manufacturing. As technologies advance, semiconductor manufacturing processes are becoming more and more complicated, and how to maintain their production yield becomes an important issue. A smart manufacturing platform can be designed to realize yield enhancement and assurance [22, 23]. Fig. 4.5 illustrates smart semiconductor manufacturing factory based on an edge computing framework [24]. It is here noted that edge computing in a manufacturing framework delivers high response, low-latency speedy results. Using a mix of technologies to leverage manufacturing productivity. Edge computing uses judiciously selected apps with real-time processing abilities to aid trained operators and sensorized technology systems to mine and capture manufacturing data sources and use these to turn actions into speedy manufacturing productivity and safety results.

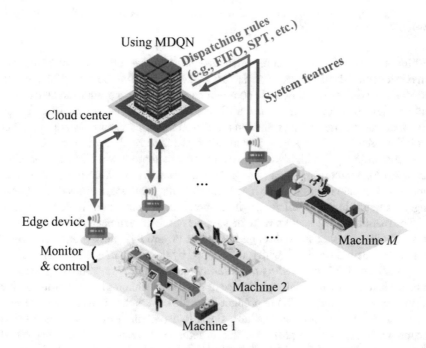

Fig. 4.5 Smart semiconductor manufacturing based on an edge computing [24]

- *Energy System*: Fuel cell vehicles utilize onboard stored hydrogen to produce onboard electricity to power the electric motors. The Department of Energy (DOE) has set some targets for onboard hydrogen storage, which includes a limit on the operating temperature and refuel rate of the hydrogen storage system. Smart manufacturing technologies can be adopted to address the operational challenges in the onboard hydrogen storage in metal hydride cells.
- *Smart Machining*. In smart manufacturing, the deployment of smart machining can be achieved with the help of smart robots and various other types of smart sensorized objects that are capable of real-time sensing. CPS-enabled smart machine tools are used for producing products. For example, CPS-enabled smart machine tools can capture the real-time data so that machine tools and their services could be synchronized to provide smart manufacturing solutions. Various sensors and data acquisition devices are deployed in the machine tools to collect real-time machining data [25].

Other applications include automotive industry, smart monitoring, smart control, the blade smart manufacturing system, smart resin transfer molding systems, and prediction of the healthiness of the smart manufacturing line. These applications are just the tip of the iceberg. New applications are likely to emerge when value chains are fully linked to smart factories.

4.8   Smart Factory

While manufacturing may be regarded as a process that turns raw materials into physical value-added products, the factory is the structure where the manufacturing occurs. In traditional factories, the communication between a product and the different operators that act on the value chain is usually slow and inefficient. There is also a lack of traceability of the product in traditional factories. Smart systems, together with the Industry 4.0 paradigm, provide alternatives to tackle the issues [26].

A smart factory is a highly digitized and connected production facility that relies on smart manufacturing. It can integrate data from system-wide physical, operational, and human assets to drive manufacturing, maintenance, and inventory tracking. This results in a more efficient, agile, responsive, proactive, and predictive system.

The manufacturing sectors with the support of the government are working hard to secure their market share through inventions toward the fourth industrial revolution known as Industrie 4.0, which was coined by the German government in 2011 [27]. *Industry 4.0 is enabled by the Internet and* designed for improving the way modern factories operate through the use of the latest technologies. According to the architecture of Industry 4.0, the product lifecycle covers the smart product design, smart production, smart manufacture, smart logistic, smart marketing, and smart maintenance and service [28]. The smart factory is a manufacturing concept that provides efficient, flexible, and adaptive production solutions in a world of increasing complexity. Currently, the smart factory is regarded as the factory of the future.

The terms "smart factory," "smart manufacturing," "Industry 4.0" and "factory of the future" represent keywords to denote what industrial production will look like in the future. They refer to a factory with a manufacturing solution that provides a flexible, adaptive, and efficient production. The smart factory refers to a leap forward from more traditional automation to a fully connected, sensorized, and flexible manufacturing system. A factory is made smart by communication technologies and the ability to use the data from connected operations. Being a smart factory also suggests an integration of shop floor decisions and insights with the rest of the supply chain. As shown in Fig. 4.6, the smart factory is a move toward a factory-of-things, which is well aligned with the Internet of things [29]. A smart factory is an intelligent factory which is capable of improving competitiveness, productivity, quality, safety, and customer satisfaction.

The structure of a smart factory can include a combination of production, information, and communication technologies. Central to the smart factory is the technology that makes data collection possible. These include the intelligent sensors, motors, and robotics. The defining features of a smart factory are connectivity, autonomy, optimization, transparency, proactivity, and agility [30].

- *Connectivity:* Smart factories require the operations, processes, and materials to be connected to generate the data necessary to make real-time decisions. The entire factory system is connected with a network of sensors, switches, motors, etc. The connected production facility relies on smart manufacturing. Each facil-

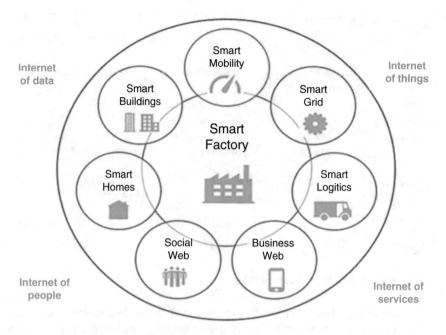

Fig. 4.6 A typical smart factory [29]

ity is linked to the others and to the entire enterprise both within the factory and beyond. Connectivity enables providing information at all industrial plant levels.

- *Autonomy:* Factory units are becoming more and more automated since it saves labor and material. The automation includes factory floors and robots working 24/7. Smart factories take this simple automation much further and are able to run without much human intervention. The Internet of things (IoT) standards (or the industrial IoT) will facilitate automation, allowing machines to communicate with machines in any sector. The connectivity and automation increase efficiency and productivity.

- *Optimization*: An optimized smart factory allows operations to be executed with minimal human intervention. Smart factory creates an environment where machinery and equipment can improve processes through automation and self-optimization.

- *Transparency:* A transparent network enables greater visibility and ensures that the organization can make more accurate decisions.

- *Proactivity:* In a proactive system, employees and systems can anticipate and act before problems arise. The ability of the smart factory to predict future outcomes can improve productivity, quality, and safety. The factory of tomorrow will be a proactive, responsive, and self-healing system.

- *Agility:* Agile flexibility allows the smart factory to adapt to schedule and production changes with minimal intervention. The agility of future smart factories in terms of a continuous reconfiguration leads to dynamically changing product traffic flows in the factory.

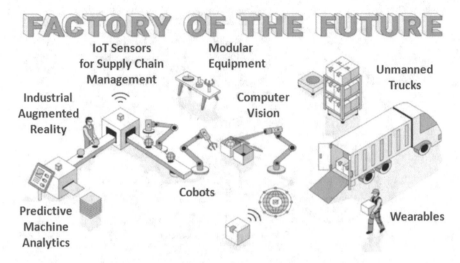

Fig. 4.7   Factory of the future (modified) [28]

The implementation of smart factory solutions makes sustainable production attainable which in turn tackles global manufacturing challenges. The essential requirements for implementing the smart factory concept include the following [31]:

- *Interoperability*: It is important to communicate efficiently using IoT.
- *Virtualization:* It is essential in order to control physical processes by CPS and create a virtual copy of the physical world. Software-defined networking (SDN) and network virtualization can help to solve the challenges of future factory networking.
- *Decentralization:* It is required due to the soaring demand for customized products, which hinders centralized control and product management.
- *Real-Time Capability*: This concerns the need for collecting and analyzing information in real time.

Figure 4.7 depicts the factory of the future [32].

4.9   Benefits

Today, manufacturing is getting smart. Smart manufacturing (SM) will transform how products are designed, fabricated, operated, and used. It has the potential to address some of the challenges facing manufacturing enterprises. Potential benefits of smart manufacturing include cost reduction, production flexibility, shorter product times-to-market, energy efficiency, environmental impact reduction, increased productivity, and quality gains. SM can revitalize the industrial sector by facilitating global competitiveness, providing sustainable jobs, radically improving performance, and facilitating manufacturing innovation. SM technologies can help manufacturers and designers to create highly differentiated, cost-effective, and competitive products that meet today's market needs while delighting the customers. SM promotes energy and environmentally efficient production. With the shortage of engineers today and the pressure to maintain a competitive edge, SM can help deliver the right products, at the right cost, at the right time to the right customers and the right locations. It can make the industry more efficient, profitable, and sustainable, reduce energy usage, produce zero incidents, and enhance faster technology and product adoption. Through increased efficiency of labor and materials, creating new jobs, and having more reliable results, most companies using SM will see an return on investment (ROI) on their investments within a short period of time [33]. With time, smart manufacturing will only get smarter, faster, and more efficient. Other benefits include the following:

- The investment of constructing a smart factory benefits manufacturers by creating a safer and more reliable plant.
- Smart factories reduce dependence on human labor and the evolving roles of employees.
- Lower costs and an enhanced customer experience. Employees should be willing to keep on developing their skills.
- Real-time visibility creates the flexibility to mobilize the supply chain to meet peak demand.
- Although smart factories produce physical goods, they also generate data and analytics as well, that could be useful in revealing the amounts of waste, and the excessive usage of electricity and water, thus paving the way for more responsible manufacturing resources management.
- Improve safety and security of the entire smart manufacturing complex.
- Increase productivity and intelligence associated with the smart manufacturing organization.

Some of these benefits are illustrated in Fig. 4.8 [34].

Fig. 4.8 Some benefits of smart manufacturing/factory [34]

4.10   Challenges

Smart manufacturing offers both benefits and challenges. Unfortunately, many manufacturing organizations are not fully aware of what SM is really about and its relevance to their organizations. Perhaps the most glaring challenge facing smart manufacturing is the risk of a cyberattack. Security must play a major role in the development of future smart manufacturing systems. A major bottleneck in smart manufacturing is achieving the communication and interaction between the physical and virtual spaces of manufacturing. SM technologies cannot be adopted independently. The workforce skills gap generated by a more-connected manufacturing system must be addressed and given a top human resources management and training priority in companies planning to adopt SM. Smart manufacturing also critically depends on the right sets of policies in order to flourish. These policies are determined over time by governmental and non-governmental organizations catering to the analysis of smart manufacturing systems information. Successful smart manufacturing requires standards for developers and users. The use of standards in the manufacturing environment is essential for creating repeatable processes and reliable systems.

Creating environmentally sustainable manufacturing is the key challenge for smart factories of the future. The high interconnectivity of the smart factory makes the impacts of critical infrastructure breakdowns to be substantial. The idea of adopting or implementing a smart factory can appear complicated or insurmountable. By starting small with manageable smart systems and components and slowly growing, the promise of the smart factory can become a reality. Some old factories cannot face challenges in keeping up with ever-shifting trends. For the full promise of smart manufacturing to be realized, these challenges must be addressed.

4.11   Global Adoption of Smart Manufacturing

Smart manufacturing has been a topic of global interest among practitioners, strategists, academicians, and governments. It has been widely recognized as a groundbreaking manufacturing trend that will revolutionize modern manufacturing industries. With the fast development of the global economy, the manufacturing industry is going through a tremendous revolution in both developed and developing nations. The manufacturing sector worldwide is increasingly becoming automated and digitized. Various governments have proposed SM initiatives to facilitate the development of their own manufacturing industries and issued their national advanced manufacturing development strategies. Many large companies from the United States, Japan, Korea, and European countries have initiated collaborative efforts on the future of manufacturing and have started to adopt SM. Small- and medium-sized enterprises (SMEs) (also known as Smart Laundry Services) are still struggling with developing an SM investment portfolio. We now consider how the following nations have adopted smart manufacturing [35–40].

- *United States:* The United States is making concerted effort to advance innovation policies that will encourage the development of new technologies. These policies promote the application of modern information technologies which drive the development of smart manufacturing. The manufacturing sector has a large footprint in the US economy. However, there is a noticeable lack of coordination among the various federal departments and agencies engaged in information governance issues. In the United States, the Smart Manufacturing Leadership Coalition (SMLC) facilitates the broad adoption of manufacturing intelligence. Ohio is regarded as the nation's third largest manufacturing state, with heavy tie with the automotive industry.
- *Germany:* Germany may be a trailblazer when it comes to smart manufacturing, but signals are mixed. The German government promotes the computerization of manufacturing industries in their "Industrie 4.0" initiative, which is increasingly affecting European policy. The national goal is to create an industrial sector with the ability to compete in the global market. While its automotive industry is its powerhouse, beefing up research and development on electric and hybrid vehicles, Germany is vulnerable to global economic uncertainty and trade tensions. Volkswagen plans on becoming the first auto company to deploy 3D printing at scale. Daimler and BMW are jointly investing heavily in ride-sharing, autonomous vehicles.
- *United Kingdom:* Manufacturing is regarded as a moneymaker for the United Kingdom. British engineering and manufacturing industries have been in the vanguard of change. Manufacturing is strategic for the future strength and resilience of the UK economy. Many UK manufacturing companies are world class. Manufacturing leads other sectors in many areas such as productivity, exports, and R&D. UK manufacturing needs to evolve in order to maintain international competitiveness and promote economic, social, and environmental sustainability. The concept of the Factory of the Future (Factory 2050) provides a focus for manufacturing research. This brings into the purview some industries such as aerospace, automotive, chemicals, and pharmaceuticals, while placing out-of-scope industries such as food, beverages, tobacco, and publishing.
- *China:* China's top-down approach stand in stark contrast to the US market-driven approach. Although China's manufacturing industry is experiencing rapid development, most companies operate at industrial levels 2.0 and 3.0. In order to enhance innovation and productivity in the manufacturing industry, Made in China 2025 and Internet plus manufacturing are issued by the Chinese government. Smart manufacturing is listed in the five major projects and nine key tasks proposed in the plan.
- *India:* Only 28% of the Indian manufacturing industry has an implemented smart factory. A key enabler of connected factories in India is the Government's Make in India initiative, which aims to develop the country as the factory of the world and create highly skilled jobs in the sector. Some question whether the Indian infrastructure is ready for smart manufacturing and whether it is equipped with

the skills required to adapt to these evolving advanced technology concepts. A real breakthrough in smart manufacturing will come when the SMEs in India start partnering with IoT platform startups that enable higher efficiency. India must invest in R&D and in a highly trained and skilled workforce.

4.12   Conclusion

A sweeping revolution in manufacturing systems is underway in many developed countries. Manufacturing is no longer just about making physical products, it is beginning to make legacy systems more intelligent or smart. Manufacturing is evolving in many developed nations to become smart, more automated, digitalized, computerized, and complex.

Smart manufacturing is here to stay and it continues to evolve. It is a key component of the broader thrust toward Industry 4.0. It is the marriage of advanced manufacturing capabilities with digital productivity, quality, safety leveraging, smart manufacturing technologies. Smart manufacturing is regarded as a transformative technology with a high degrees of sensorized and intelligent automation. Competitiveness considerations demand participation in manufacturing planning and control innovations. Smart factories link sensorized digital technologies with smart production processes. Manufacturers today are characterized by the ability to fulfill orders on demand by doing business through short-term networks where they negotiate value-adding processes dynamically.

Modern manufacturing systems require timely and efficient production tasks. Almost all aspects of manufacturing will need to change as the industry transitions to become fully digital. Smart manufacturing systems will ultimately transform product manufacturing, customization, and delivery. They can optimize the entire business processes and operation procedure of manufacturing with the goal of achieving higher levels of productivity and quality gains. In order to retain and attract workers with the right skills, to support smart manufacturing, manufacturers are providing leadership training, advanced technology skills training, and employee performance management immersion exposures.

The future of first rate-smart manufacturing is indissolubly bound with the smart factory innovation. Companies that are slow to adopt SM technologies could be left behind. More information on smart manufacturing and the smart factory can be found in the books in [8, 41–43] and the following related journals and magazines:

- *Journal of Engineering Manufacture*
- *Journal of Manufacturing Systems*
- *Journal of Intelligent Manufacturing*
- *Journal of Manufacturing Technology Management*
- *Manufacturing Letters*
- *Smart Manufacturing Magazine*
- *The Manufacturer Magazine*
- *Manufacturing Global*

References

1. N. Anwer, B. Eynard, L. Qiao (eds.), Editorial for the special issue on 'smart manufacturing and digital factory'. J. Eng. Manuf. **233**(5), 1341 (2019)
2. M.N.O. Sadiku, O.D. Olaleye, S.M. Musa, Smart manufacturing: A primer. Int. J. Trend Res. Develop. **6**(6), 9–12 (2019)
3. Y. Nukman et al., A strategic development of green manufacturing index (GMI) topology concerning the environmental impacts. Proc. Eng. **184**, 370–380 (2017)
4. D. Swathisri, D.S.S. Kumar, Green manufacturing technologies – A review (n.d.), https://www.researchgate.net/publication/305731124_GREEN_MANUFACTURING_TECHNOLOGIES_-_A_REVIEW
5. https://www.engr.uky.edu/ism
6. G.D. Maruthi, R. Rashmi, Green manufacturing: It's tools and techniques that can be implemented in manufacturing sectors. Mater. Today: Proc. **2**, 3350–3355 (2015)
7. M.N.O. Sadiku, *Emerging Smart Technologies* (Arthur House, Bloomington, 2021), p. xi
8. T.F. Edgar, E.N. Pistikopoulos, Smart manufacturing and energy systems. Comput. Chem. Eng. **114**, 130–144 (2018)
9. F. Tao et al., Digital twins and cyber–physical systems toward smart manufacturing and industry 4.0: Correlation and comparison. Engineering **5**(4), 653–661 (2019)
10. 10 Numbers on smart manufacturing (n.d.), https://blog.criticalmanufacturing.com/10-numbers-on-smart-manufacturing/
11. S. Jain. D. Lechevalier, and A. Narayana, Towards smart manufacturing with virtual factory and data analytics, in Proceedings of the 2017 Winter Simulation Conference, 2017
12. P.P. Khargonekar, Future of smart manufacturing in a global economy (2018), http://faculty.sites.uci.edu/khargonekar/files/2018/05/Smart_MFG_1.0.pdf
13. A. Kusiak, Smart manufacturing. Int. J. Prod. Res. **56**(1–2), 508–517 (2018)
14. Smart Manufacturing (n.d.), https://grantek.com/capabilities/smart-manufacturing?utm_source=google&utm_medium=cpc&utm_campaign=smart_manufacturing&gclid=EAIaIQobChMIj92SjZq65QIVgobACh0h2ATwEAMYASAAEgIo_vD_BwE
15. T.C. Chan, The development of smart manufacturing and cases study in Taiwan, in Proceedings of IEEE International Conference on Advanced Manufacturing, 2018, pp. 117–118
16. H.S. Kang et al., Smart manufacturing: Past research, present findings, and future directions. Int. J. Precis. Eng. Manuf. Green Technol. **3**(1), 111–128 (2016)
17. G.S. Ogumerem, E.N. Pistikopoulo, Smart manufacturing, in *Kirk-Othmer Encyclopedia* (Wiley, 2014
18. Smart Manufacturing – The Landscape Explained, White Paper #52, A MESA International white paper (2016), http://www.mesa.org/en/resources/MESAWhitePaper52-SmartManufacturing-LandscapeExplainedShortVersion.pdf
19. P. O'Donova et al., An industrial big data pipeline for data-driven analytics maintenance applications in large-scale smart manufacturing facilities. J. Big Data **2** (2015)
20. L. Ogbeveon, Smart manufacturing: The rise of the machines (2019), https://oden.io/blog/smart-manufacturing-the-rise-of-the-machines/
21. Manufacturing operations management talk (n.d.), https://www.manufacturing-operations-management.com/manufacturing/2016/01/on-the-journey-to-a-smart-manufacturing-revolution.html
22. Y.C. Lin et al., Development of advanced manufacturing cloud of things (amcot – A smart manufacturing platform). IEEE Robot. Autom. Lett. **2**(3), 1809–1816 (2017)
23. J. Moyne, J. Iskandar, Big data analytics for smart manufacturing: Case studies in semiconductor manufacturing. PRO **5** (2017)
24. C.V. Lin et al., Smart manufacturing scheduling with edge computing using multiclass deep Q network. IEEE Trans. Ind. Inf. **15**(7), 4270–4284 (2019)
25. P. Zheng et al., Smart manufacturing systems for industry 4.0: A conceptual framework, scenarios and future perspective. Front. Mech. Eng. (2018)

26. T.M. Fernández-Caramés, P. Fraga-Lamas, A review on human-centered IoT-connected smart labels for the industry 4.0. IEEE Access **6**, 25939–25957 (2018)
27. M.M. Mabkhot et al., Requirements of the smart factory system: A survey and perspective. Mach. Des. **6**(2) (2018)
28. H. Zhao, J. Hou, Design concerns for industrial big data system in the smart factory domain: From product lifecycle view, in Proceedings of the 23rd International Conference on Engineering of Complex Computer Systems (2018), pp. 217–220
29. What are the benefits of a smart factory? (n.d.), https://www.quora.com/What-are-the-benefits-of-a-smart-factory
30. Five key features of the smart factory (n.d.), https://www.welcome.ai/news_info/research-news-information/five-key-characteristics-of-a-smart-factory
31. F. Odważny, O. Szymańska, P. Cyplik, Smart factory: The requirements for implementation of the industry 4.0 solutions in FMCG environment – Case study. Sci. J. Logist. **14**(2), 257–267 (2018)
32. Smart factory Industry 4.0 (n.d.), https://bestbarcodeworld.com/smart-factory/
33. What's so smart about smart manufacturing? (2018), https://professional.mit.edu/news/news-listing/whats-so-smart-about-smart-manufacturing
34. Industry 4.0, the smart factory & the digital supply chain (n.d.), https://www.datexcorp.com/industry-4-0-smart-factory-digitization-supply-chain/
35. S. Mittal et al., A smart manufacturing adoption framework for SMEs. Int. J. Prod. Res. (2014)
36. K.D. Thoben, S. Wiesner, T. Wuest, Industrie 4.0' and smart manufacturing – A review of research issues and application examples, submitted to *Int. J. Autom. Technol.*
37. K. Ridgway et al., The factory of the future (2013), https://www.researchgate.net/publication/322788326
38. F. Tao, Q. Qi, New IT driven service-oriented smart manufacturing: Framework and characteristics. IEEE Trans. Syst. Man Cybern. Syst. **49**(1), 81–91 (2019)
39. Smart factories of India (n.d.), https://www.exito-e.com/smart-factories-of-india/
40. B. Buntz, Is my smart factory smarter than yours? It's hard to say (2019), https://www.iot-worldtoday.com/2019/05/30/is-my-smart-factory-smarter-than-yours-its-hard-to-say/
41. F. Tao, M. Zhang, A.Y.C. Nee, *Digital Twin Driven Smart Manufacturing* (Academic Press, 2014)
42. M. Soroush, M.M. Baldea, T. Edgar (eds.), *Smart Manufacturing: Concepts and Methods* (Elsevier, 2020)
43. A. Arockiarajan, M. Duraiselvam, R. Raju, *Advances in Industrial Automation and Smart Manufacturing* (Springer, 2021)

Chapter 5
Industry 4.0

The Fourth Industrial Revolution is not about new Apps or new technologies. It is about a new era, new ways of thinking and new ways of doing business.

–Nicky Verd.

5.1 Introduction

The manufacturing industry is crucial to any nation's economy because it profoundly impacts economic and societal progress. It is the main driver of research, innovation, productivity, wealth creation, job opportunities, and export that supports a favorable international balance of trade position and the prestige and business advantages that go along with it. Manufacturing companies in all nations are facing an urgent need to improve quality, boost productivity, lower cost, and have quick customer services. To meet these challenges, the manufacturing industry needs to promote technological innovation and complete manufacturing, business, technological, and military intelligence upgrading.

The traditional manufacturing system basically includes two major parts: humans and physical systems. Manufacturing technology progressed from Henry Ford's assembly line to computerization. Gradually, physical systems replace humans by taking over the majority of work. The transformation of the manufacturing industry from traditional manufacturing to new-generation manufacturing is a process of evolution [1]. Today, most of the manufacturing systems are changing rapidly in terms of the adoption of new, emerging technologies. Emerging technologies have raised the levels of digitalization, networking, and intelligence in the manufacturing industry. Recently, industrial technology has undergone profound changes, leading to "Industry 4.0" era.

Industry 4.0 (or I4.0) has been introduced recently as the trend toward digitization and automation of the manufacturing sector. It is the convergence of digital and physical technologies disrupting the manufacturing industry, which is resulting in the blurring of boundaries between the physical, digital, and biological worlds. Industry 4.0 is the convergence of nine digital industrial technologies: advanced robotics, additive manufacturing, augmented reality, simulation, horizontal/vertical integration, cloud, cybersecurity, big data analytics, and the industrial Internet [2, 3].

M. N. O. Sadiku et al., *Emerging Technologies in Manufacturing*, https://doi.org/10.1007/978-3-031-23156-8_5

The term 4IR, coined in 2016 by Klaus Schwab, founder and executive chairman of the world economic forum, author of the book *Fourth Industrial Revolution*, is preceded by the German Government sponsorship of "Industrie 4.0" project to promote the computerization of manufacturing in 2011, which term is now synonymous with the term Industry 4.0 [4, 5]. *It focuses on creating intelligent or smart environment within the production system.* For essential emphasis, the emerging technologies giving rise to Industry 4.0 include cyber-physical systems, Internet of Things, cloud computing, augmented reality, edge computing, big data, predictive analytics, intelligent robots, 3D printing, genetic engineering, artificial intelligence, industrial integration, and other contemporary and contributory revolutionary technologies. These technologies are poised to transform manufacturing around the globe.

Although Industry 4.0 is a manufacturing initiative, it has been involved in smart transportation and logistics, smart buildings, oil and gas, smart healthcare, and smart cities, to name a few. This trend of the Industry 4.0 affects people and processes throughout the society, redefining family life, globalization, markets, etc. Industry 4.0 enables the digitalization of the manufacturing sector with built-in sensing devices virtually in all manufacturing components, products, and equipment. It is the current trend in automation and data exchange in manufacturing organizations [2].

This chapter provides a brief introduction to Industry 4.0. It begins by covering the four industrial revolutions leading to Industry 4.0. It provides an overview on the concept of Industry 4.0. It describes the design principles and characteristic features of Industry 4.0.

It gives some of the applications of Industry 4.0. It highlights some benefits and challenges facing Industry 4.0. It covers the global adoption of I4.0. The last section concludes with comments.

5.2   Industrial Revolutions

The following industrial revolutions led to Industry 4.0 [6, 7].

- *The first industrial revolution:* The revolution happened between the late 1700s and early 1800s. It witnessed the transition from handmade agricultural production to mechanical large-scale factory production in the nineteenth century.
- *The second industrial revolution*: The revolution is also known as the Technological Revolution and spanned the period from the 1850s to World War I and saw the age of steam power, electrification of factories, and new manufacturing "inventions." The new manufacturing enabled assembly and led to mass production and, to some degree, the introduction of automation in the manufacturing industry.
- *The third industrial revolution:* This revolution is also known as the Digital Revolution and took place from the late 1950s to late 1970s and saw the age of computerization, a change from analog technology to digital technology. This

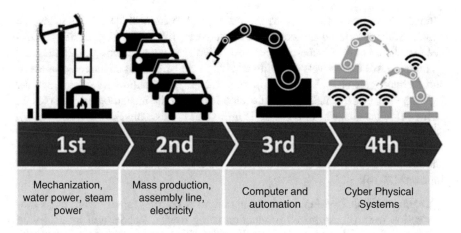

Fig. 5.1 Four stages of industrial revolution [8]

revolution has continuously pushed industrial development to a new level. It had to do with the rise of computers, computer networks (WAN, LAN, MAN), the rise of robotics in manufacturing, connectivity, and the birth of the Internet.

- *The fourth industrial revolution:* Finally, this revolution starting circa 2000–2016, or a few decades earlier, having been coined in Germany in 2011 at the Hannover Fair event as "Industrie 4.0" and having been referenced as the fourth industrial revolution (4IR) in 2016 by Klaus Schwab, is marked by full digitization and automation of manufacturing processes and is happening right now. It moves the pace of technological development from the Internet and the client-server model to ubiquitous technological fusion and mobility, the bridging of digital and physical environments, to the convergence of information technology (IT) and operational technology (OT) and emerging technologies such Internet of Things, big data, cloud computing, advanced robotics, etc.

The four (4) sequential revolutions have resulted in radical changes in manufacturing. They are illustrated in Fig. 5.1 [8]. The fourth industrial revolution is commonly referred to as Industry 4.0. Industry 4.0 is about the information-intensive digital transformation of industry and manufacturing. Thus, Industry 4.0 is variously known as the "Fourth Industrial Revolution," "smart manufacturing," or "industrial Internet."

5.3   Overview on Industry 4.0

The term "Industry 4.0" came from the German term "Industrie 4.0," which was first used in 2011 in a project sponsored by the German government that was meant to promote the computerization of manufacturing [4]. To meet the challenges of traditional manufacturing, Germany introduced the concept of "Industry 4.0"; the

Chinese government put forward the concept of "China Made 2025," while the United States proposed the concept of "industrial Internet" [9]. Industry 4.0 has been predicted to be a game-changer, revolutionizing manufacturing practices.

Industry 4.0 collectively refers to a wide range of current concepts, processes, and technologies [10]. As shown in Fig. 5.2 [11], the required technologies for Industry 4.0 transformation include cloud computing technologies, Internet of Things (IoT), industrial Internet of Things (IIoT), cyber physical systems (CPS), advanced materials, additive manufacturing, cloud computing, artificial intelligence, machine learning, cybersecurity, big data analytics, cognitive computing, autonomous robots, and mobile services. Some of these are briefly explained as follows [12–14]:

- *Intelligent Robots:* Manufacturing paradigm is shifting rapidly from mass production toward customized production using robots. Robots will interact with one another and work safely side by side with humans. Robots will be more

Fig. 5.2 Key technologies for Industry 4.0 [11]

dominant in manufacturing and will dominate the production process. With the deployment of many intelligent industrial robots working as a team, it is feasible to have a larger range of manufacturing applications.

- *Internet of Things:* IoT refers to a network of physical devices that are digitally interconnected and facilitate communication and exchange of data through the Internet. It allows people and things to be connected anytime, anyplace, with anything and anyone. IoT enables a ubiquitous presence for a common purpose of various things or objects interacting and cooperating with each other, digitalizing all physical systems. Industry 4.0 means that more devices will be enriched with embedded computing. This will allow field devices to communicate and interact both with one another. The Internet of Things is complemented by Internet of Services. The industrial IoT (or IIoT) is the implementation of IoT in the industrial sector. This is one of the important backbones of Industry 4.0.
- *Cybersecurity:* Devices are increasingly connected to the global network, the Internet. The possible damage of cyberattacks is substantial in terms of continuity of business operations, theft of confidential information, and reputational harm. With the increased connectivity associated with Industry 4.0, there is need to protect critical industrial systems and manufacturing lines from cyber threats.
- *Cloud Computing:* Cloud computing is a means of pooling and sharing hardware and software resources on a massive scale. Users and businesses can access applications from anywhere in the world at any time. The main objective of cloud computing is to make a better use of distributed resources and solve large-scale computation problems. If used properly, cloud computing is a technology with great opportunity for businesses of all sizes. The adoption of cloud computing has several advantages related to cost reduction.
- *Big Data:* Data is becoming more and more accessible and ubiquitous in many industries, leading to the issue of big data. Big data usually comes from various sources such as sensors, devices, video/audio, networks, assorted generalized, and/or environmental databank systems and social media. In an Industry 4.0 system or environment, the collection and comprehensive evaluation of data from many different sources will become as indispensable, and it is challenging. Big data is characterized in terms of ten (10) Vs: volume, variety, velocity, veracity, vision, volatility, verification, validation, variability, and value. Manufacturing competitiveness in the international market is enhanced by these characteristics.
- *Additive Manufacturing (AM):* This is the process of creating a 3D object based on the deposition of materials on a layer-by-layer or drop-by-drop basis. AM technologies are commonly referred to as rapid prototyping, layered manufacturing, digital manufacturing, or 3D printing. With Industry 4.0, these additive-manufacturing methods will be widely used to produce small batches of customized products.
- *Augmented Reality (AR):* Increasing human performance is the main aim of AR. AR technology is applied in a wide range of sectors, including entertainments, marketing, tourism, surgery, logistics, manufacturing, education, and maintenance. Companies will make much broader use of augmented reality to

provide workers with real-time information to improve decision-making and work procedures.

- *Automation:* This involves using control systems to operate a process or system with minimal or no direct human intervention. Industry 4.0 can be defined as "a name for the current trend of automation and data exchange in manufacturing technologies." Autonomous systems make decisions on their own. The automation increases efficiency and productivity. Industry 4.0 refers to the current trend of automation. This includes using machine-to-machine and Internet of Things (IoT) deployments to help manufacturers provide increased automation, improved communication, and monitoring leading to increased productivity, efficiency, waste reduction, and enhanced profits.
- *Cyber-Physical System:* CPS is the merger of "cyber" as in electrical and electronic systems with "physical" things. A CPS is a mechanism through which physical objects and software are closely intertwined. It consists of physical entities (such as machines, vehicles, work pieces, etc.), which are equipped with technologies such as RFIDs, sensors, microprocessors, telematics, or embedded systems. I4.0 can be assumed as a CPS production.
- *Simulation:* Computer simulation is an indispensable and powerful tool for successful implementation of the digital manufacturing system. Simulation modelling involves using models of a system or a process to better understand or predict and improve the behavior or performance of the system or process. Like in experimentation, simulation modeling allows for the validation of products, processes, systems, and configurations in very short time orders such as milliseconds or microseconds. It helps with cost reduction, decreases development cycles, and increases product quality. It allows predictive analysts and simulationists to gain insights into complex systems and tests before their real implementation.

Industry 4.0 may be regarded as the convergence and applications of the above simulation-based digital technologies. These are some of the important technologies transforming industrial production. These technologies will help Industry 4.0 to realize smart industry and smart manufacturing goals.

5.4   Principles and Features

This section considers the design principles and key characteristics of Internet 4.0, popularly known as Industry 4.0. As shown in Fig. 5.3, there are six design principles in Industry 4.0 [15, 16].

- *Interconnection:* The ability of machines, devices, and humans to connect and communicate with each other via the Internet of Things (IoT). This requires collaboration, security, and standards.

Fig. 5.3 Six design principles in Industry 4.0 (modified) [15]

- *Information Transparency:* The ability of information systems to create a virtual copy of the physical world by enriching digital plant models with sensor data. This includes data analysis and provision of information.
- *Technical Assistance*: This refers to the ability of CPS to physically support humans by conducting a range of tasks that are unpleasant, too exhausting, or unsafe for their human coworkers. This requires visual and physical assistance.
- *Decentralized Decisions*: The ability of CPS to make decisions on their own and to perform their tasks as autonomously as possible. Their tasks are delegated to a higher level only when there is interference or conflicting goals.
- *Interoperability:* This is also known as collaboration. It is the ability to have many standards communicate with or talk to each other and exchange data. Interoperability means connected devices, connected communication technologies, connected people, and connected data that are collaborating to achieve the manufacturing goals and objectives of the enterprise. The maturity of cyber-physical systems allows humans and factories to connect and communicate with each other and derive insights in real time in support of increased efficiency and enhanced productivity and profits for the organization(s).

- *Modularity:* This has to do with a shift from rigid systems, inflexible models, and planning to an environment where changing demands from customers, partners, and regulators market conditions are monitored, analyzed, and utilized causing the need for transformation and flexibility leading to increased efficiency, productivity, and profits.

The main characteristic features of Industry 4.0 include the following [17]:

- *Decentralization:* Decentralization in Industry 4.0 organizations refers to a shift away from centralized control systems to decentralized control.
- *Flexibility*: This is a key attribute of Industry 4.0. It refers to the capability to position manufacturing to respond to changing demands. It is enabled by the application of cyber-physical production systems. Industry 4.0 allows a high flexibility in the operation of automated systems.
- *Interconnectedness:* This is a basic principle of Industry 4.0. It enables the communication and information exchange between entities in the system. IoT enables interconnectedness.
- *Customization:* The mass customization of production, products, and services is regarded as a fundamental paradigm shift of Industry 4.0. This allows individual preferences to be included in the design in a cost-effective manner and is designed to increase users' participation.
- *Efficiency:* This is an important consequence in the Industry 4.0 deployment or realization and is made possible by its enabling technologies. Industry 4.0 will ensure factories become smart and adaptable, leading to an improvement in their resource efficiency.
- *Integration*: This essentially refers to the enterprise-wide integration of enabling technologies and to a slightly lesser extent also to the value chain and the process level contributions to the manufacturing or production functions. In Industry 4.0 context, there are three types of integration: horizontal integration, vertical integration, and end-to-end integration. The vertical integration is about the integration of IT systems at various hierarchical production and manufacturing levels. It will make the traditional automation pyramid view disappear. The horizontal integration is about the integration of OT systems and combines the operational technology (OT) to support the end-to-end value chain: from supplier and ultimately the customer. Horizontal integration helps with horizontal coordination, collaboration, cost savings, value creation, and speed.

5.5 Applications

Industry 4.0 is still in the early stages of implementation. One of the original objectives of Industry 4.0 was serving small and medium enterprises (SME) first. The major applications of Industry 4.0 are smart factory, smart/intelligent manufacturing, smart product, and smart city [18].

- *Smart Factory:* This is where the Internet, wireless sensors, and other advanced technologies work together to optimize the production process and improve customer satisfaction. In smart factories, humans, machines, and resources communicate with each other, like in a social network. Industry 4.0 is an emerging network approach where components, processes, and machines are becoming smart. Factories will gradually become automated and self-monitoring as the machines are given the ability to communicate with each other and their human coworkers. The smart factory has overcome vendor-specific, stand-alone solutions and is creating a solid base for cross-vendor solutions within the manufacturing environment. Manufacturing will completely be equipped with sensors, actuators, and autonomous systems and will be decentralized. Figure 5.4 illustrates some elements of a smart factory [19].
- *Intelligent Manufacturing:* This is also known as smart manufacturing. It plays a crucial role in Industry 4.0. It is similar to cloud manufacturing and IoT-enabled manufacturing. It takes advantage of advanced information and manufacturing technologies to achieve flexible, smart, and reconfigurable manufacturing processes in order to address the dynamics and fluctuations of the global market. It requires some underpinning technologies in order to enable devices to vary their behaviors in response to various situations. Intelligent manufacturing entails intelligent products, intelligent production, intelligent services, etc. It is not yet mature but is emerging.
- *Predictive Maintenance:* Due to the use of IoT sensors, Industry 4.0 technology enables manufacturers to predict when potential problems will arise before they actually happen. Predictive maintenance can identify maintenance problems with dependable predictability and high precision. It enables manufacturers to pivot from preventive maintenance to predictive maintenance. With IoT systems in place, preventive maintenance is automated and streamlined. Manufacturers

Fig. 5.4 Elements of a smart factory [19]

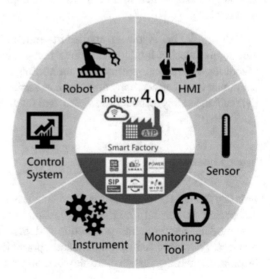

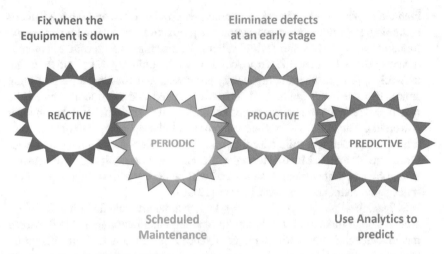

Fig. 5.5 Predictive analytics [20]

that can employ predictive maintenance and make automated predictive decisions will improve productivity and gain a competitive advantage. Fig. 5.5 depicts predictive analytics [20].

- *Mining Industry:* Current challenges facing the mining industry include high global demand for raw materials, strict labor market, high standards of production requirements, saving natural resources, international competition, and continued declining commodity prices and daily operational and industrial accident risks associated with mining operations and the attendant liability implications of those risks. To address these challenges, it is imperative to have a real-time flow of information between ERP. Industry 4.0 architecture and the allied technologies are applied to address the challenges that exist in the mining environment [21].

Industry 4.0 has also been applied in semiconductor industry, furniture industry, aluminum industry, food industry, automotive industry, process optimization, manufacturing operations, and quality monitoring. Those who promote Industry 4.0 claim that it will affect many areas such as services and business models, productivity, machine safety, product lifecycles, and industry value chains.

5.6 Benefits

Industry 4.0 ushers manufacturing and business actors, customers, the government, and society in general into the digitalization era, where everything is digital: business models, environments, production systems, machines, operators, products, and services. The rise of Industry 4.0 makes it possible to gather and analyze data across machines, enabling faster, more flexible, and more efficient processes to produce higher-quality goods at reduced costs. Industry 4.0 is revolutionizing manufacturing, increasing flexibility, improving production processes, increasing productivity, affecting the whole product lifecycle, creating new business models, changing the work environment, and restructuring the labor market. It will also cause macroeconomic shifts, foster industrial growth, increase production speeds, and boost employment. It benefits the factory, business, products, customers, and society in general [11]. It has potential to positively affect meeting individual customer requirements, production flexibility, data-driven decision-making, better customer proximity, efficiency, value creation opportunities through new services, and a competitive economy. The capabilities of Industry 4.0 lead to the "smart anything" phenomena which include smart grid, smart energy, smart cities, smart logistics, smart facilities, smart buildings, smart plants, smart manufacturing, smart factories, smart mining, smart oil and gas, smart healthcare, smart education, and smart "x" or "anything."

Other benefits include [15]:

- Enhanced productivity through optimization and automation.
- Real-time data for a real-time supply chain in a real-time economy.
- Higher business continuity through advanced maintenance and monitoring possibilities.
- Better quality products: real-time monitoring, IoT-enabled quality improvement predicated on the cooperation of humans and robots in an enclosed workspace, the robots otherwise being termed "cobots".
- Better working conditions and sustainability.
- Personalization and customization for the "new" consumer.
- Improved agility.
- Development of innovative capabilities and new revenue models.
- Increasing number of vertical industries that are adopting Industry 4.0.
- Makes manufacturing and business actors, customers, in general more competitive, especially against business and industrial giant disruptors such as Amazon, Facebook, Alibaba, etc.
- Gain of competitive advantages in domestic and global markets.
- Makes manufacturing and business teams stronger and more collaborative.
- Enhance computational power and connectivity.
- Promote human-machine interaction (cobots).
- Focus on analytics and intelligence.
- Encourage advanced production methods.

5.7 Challenges

Companies face formidable challenges in the adoption of Industry 4.0 because it is hard to achieve Industry 4.0. The challenges faced in the implementation of Industry 4.0 include [8]:

- Unclear legal issues and data security issues.
- Threat of redundancy of the corporate IT department.
- Loss of many jobs to automated manufacturing and industrial processes.
- Reliability and stability needed for critical machine-to-machine operations and the supporting information technology (IT) and operational technology (OT) communication needed.
- Need to maintain the integrity and efficiency of operation and the predictability of the production processes.
- Lack of adequate skillsets to expedite the march toward the fourth industrial revolution.
- General reluctance to change by stakeholders.
- Unclear economic benefits in relation to the significant or, rather, excessive investment required.
- Lack of regulations and standards, with respect to the various forms of certifications required.
- Heterogeneity of systems involved and the challenges of integrating heterogenous data from various sources.
- Increasing age of employees with implications for agility, retraining, wage structure, and the adverse effects of globalization of markets on the American consumers.
- New generation of machines needed, with implications for huge investments in trendy mechanization and automation to command participation in Industry 4.0.
- No common understanding in terms of assessing the Industry 4.0 readiness for organizations.

Some of these challenges are illustrated in Fig. 5.6 [22]. All contributing parties collaborate well to overcome the challenges mentioned above. It is crucial for manufacturing organizations to self-assess their Industry 4.0 readiness to survive and thrive in the age of Industry 4.0. Industry 4.0 readiness refers to the degree to which manufacturing companies can take advantage of Industry 4.0 technologies. Manufacturing organizations can improve their competitive advantage in domestic and global market by making significant investments in improving their readiness toward Industry 4.0 technologies. This will help them to have an active and greatly rewarding role in future global markets [23].

Fig. 5.6 Some challenges with Industry 4.0 [22]

5.8 Global Adoption of Industry 4.0

The techniques of Industry 4.0 have gone global. Industry 4.0, also known as the fourth industrial revolution, has caught the attention of organizations worldwide. It has become the future of global manufacturing. Nations around the world are actively upgrading their manufacturing industries and engaging in the adoption of Industry 4.0. Businesses everywhere can benefit from embracing Industry 4.0. This will create wealth and plenty of high technology job opportunities on the one hand as well as some displaced low-skill workers on the other hand, who may become unemployed and who may be untrainable, leading to the wider social divide, inequity, and societal tension, granted the disruptive nature of Industry 4.0. The global adoption of Industry 4.0 is not just a matter of government initiatives but also a result of an increasing focus and participation of industrial giants. We now consider the adoption of Industry 4.0 in different nations [24, 25].

- *United States:* The United States has the Smart Manufacturing Leadership Coalition (SMLC), a nonprofit organization consisting of manufacturers, government agencies, universities, and laboratories that have the common interest of advancing the way of thinking behind Industry 4.0. It is hoped that the coalition will enable manufacturers gain affordable access to technologies that can be adopted to meet their needs. The United States has also launched the Advanced Manufacturing Partnership.
- *Germany*: Germany introduced the phenomenon of Industry 4.0 in 2011 at the Hannover Fair event. It is a strategic initiative introduced by the German government with the aim of transforming the manufacturing industry through digitalization. German manufacturing strategy played a key role in maintaining and promoting its importance as a "forerunner" in Industry 4.0. Extensive efforts were made by the European manufacturing researchers and companies to embrace it. The German government along with its companies, universities, and research institutions are developing fully automated, Internet-based "smart" factories.
- *China:* In 2015, China's government launched two actions simultaneously, i.e., the "Internet Plus" and "Made in China." The development of Industry 4.0 in China will be directly related to the development of China's industrial economy. The government announced elimination of rules that required car manufacturers such as General Motors to collaborate with a local company to open factories in China. The Chinese Ministry of Industry and Information Technology (MIIT) in China led the creation of the "Made in China" initiative. This initiative aims include: increase innovative capability in national manufacturing, boost Chinese quality brand-building, promote environmentally friendly manufacturing, enable breakthroughs in key sectors, press further restructuring of the manufacturing industry, and increase international involvement in manufacturing.
- *Japan:* Japan has a Society 5.0 strategy. In 2015, Japan started its Industrial Value Chain Initiative (IVI), which corresponds to Germany's Industry 4.0 initiative. Thirty (30) Japanese companies, including Mitsubishi Electric, Fujitsu,

Nissan Motor, and Panasonic, are part of the initiative. The IVI is a forum designed to combine manufacturing and information technologies and create an environment where enterprises can collaborate. The forum is based on two principles: connected manufacturing and the loosely defined standard.

5.9   Conclusion

Industry 4.0 is a broad domain that includes production processes, efficiency, data management, relationship with consumers, competitiveness, etc. The concept has attracted much attention all around the globe. However, Industry 4.0 is still in a conceptual state, with developed countries and Germany in particular leading the way.

Industry 4.0 may benefit emerging economies such as India. Today, the fourth industrial revolution is transforming economies, jobs, and society itself. Like any revolution, Industry 4.0 is disruptive and is rapidly changing the industrial landscape. It will have an important influence on the complete transformation and digitization of the manufacturing industry. Industry 4.0 is the current gradual industrial transformation with automation, cloud computing, cyber-physical systems, robots, and industrial IoT to realize smart industry and manufacturing. It is the new manufacturing objectives and, with the aim of achieving ideally, zero-defects production in the manufacturing industry.

The world of manufacturing is changing rapidly. In order to survive and thrive, every manufacturer must be willing to invest in Industry 4.0. In other words, to stay Industry 4.0 competitive and remain relevant, manufacturers must commit to doing four things: identifying crucial business needs, investing in technology that will meet them, building organizational capabilities, and actively adapting effective and efficient processes and synergistic cultures [26].

The Industry 4.0 era requires engineering roles with different knowledge and skills that combine IT and production knowledge. Academic institutions along with their engineering colleges can play a vital role in meeting this need. Some universities are already making a concerted effort to train the next generation of engineers that are ready to work in an Industry 4.0 environment. More information about Industry 4.0 can be found in [5, 27–32] and the following related journal: *International Journal of Advanced Manufacturing Technology* and *Journal of Intelligent Manufacturing*.

References

1. Z. Li et al., Toward new-generation intelligent manufacturing. Engineering **4**(1), 11–20 (2018)
2. M.N.O. Sadiku, S.M. Musa, O.M. Musa, The essence of industry 4.0. IJRTEM **2**(9), 64–67 (2018)

3. When was the 4th industrial revolution ?. Wikipedia, https://www.salesforce.com/blog/what-is-the-fourth-industrial-revolution-4ir/. 27 Oct 2020
4. R. Drath, A. Horch, Industrie 4.0: hit or hype? IEEE Ind Electron Mag **8**, 56–58 (2014)
5. K. Schwab, *The fourth industrial revolution* (Currency, 2017)
6. M. Moore, What is Industry 4.0? Everything you need to know. https://www.techradar.com/news/what-is-industry-40-everything-you-need-to-know
7. Fourth industrial revolution, *Wikipedia*, the free encyclopedia. https://en.wikipedia.org/wiki/Fourth_Industrial_Revolution
8. Industry 4.0, *Wikipedia*, the free encyclopedia. https://en.wikipedia.org/wiki/Industry_4.0
9. L. Wang, J. He, S. Xu, *The application of industry 4.0 in customized furniture manufacturing industry*, vol 100 (MATEC Web of Conferences, 2017)
10. H. Lasie et al., Industry 4.0. Bus Inf Syst Eng **4**, 239–242 (2014)
11. A. Luque et al., State of the industry 4.0 in the Andalusian food sector. Procedia Manuf **13**, 1199–1205 (2017)
12. Embracing Industry 4.0—and rediscovering growth. https://www.bcg.com/en-us/capabilities/operations/embracing-industry-4.0-rediscovering-growth.aspx
13. M.N.O. Sadiku, *Emerging internet-based technologies* (CRC Press, Boca Raton, FL, 2019)
14. V. Alcácer, V. Cruz-Machado, Scanning the Industry 4.0: A literature review on technologies for manufacturing systems. Eng Sci Technol Int J **22**(3), 899–919 (2019)
15. Industry 4.0: The fourth industrial revolution – Guide to Industrie 4.0. https://www.i-scoop.eu/industry-4-0/
16. T.K. Sung, Industry 4.0: a Korea perspective. Technol Forecast Soc Change **132**, 40 (2018)
17. G. Beier et al., Industry 4.0: how it is defined from a sociotechnical perspective and how much sustainability it includes – a literature review. J Clean Prod **259**, 120856 (2020)
18. Y. Lu, Industry 4.0: a survey on technologies, applications and open research issues. J Ind Inf Integr **6**, 1–10 (2017)
19. D. Turbide, Why is the smart factory concept key for manufacturing? https://www.daveturbide.com/why-is-the-smart-factory-concept-key-for-manufacturing/
20. Preventive maintenance is improved through predictive analytics. https://www.designnews.com/industrial-machinery/preventive-maintenance-impr
21. M. Sishi, A. Telukdarie, Implementation of Industry 4.0 technologies in the mining industry – A case study. Int J Min Miner Eng **11**(1), 1 (2020)
22. J. Qin, Y. Liua, R. Grosvenor, A categorical framework of manufacturing for industry 4.0 and beyond. Procedia CIRP **52**, 173–178 (2016)
23. Challenges with Industry 4.0. https://stefanini.com/en/trends/news/the-fourth-industrial-revolution-industry-4-0-challenges-and-opp
24. M.A. Soomro, M. Hizam-Hanafiah, L. Abdullah, Industry 4.0 readiness models: a systematic literature review of model dimensions. Information **11**(7), 364 (2020)
25. R.Y. Zhong et al., Intelligent manufacturing in the context of industry 4.0: a review. Engineering **3**(5), 616–630 (2017)
26. M. Boggess, 10 Trends that will dominate manufacturing in 2019. https://insights.bcmonepartners.com/view/content/H9sB0
27. V. Jirkovský, M. Obitko, V. Jirkovský, *Enabling semantics within industry 4.0* (Springer, 2017)
28. T. Devezas, J. Leitão, A. Sarygulov (eds.), *Industry 4.0: entrepreneurship and structural change in the new digital landscape* (Springer, 2017)
29. A. Ustundag, E. Cevikcan, *Industry 4.0: Managing The Digital Transformation* (Springer, 2018)
30. A. Gilchrist, *Industry 4.0: the industrial internet of things* (Apress, 2016)
31. K.K. Pabbathi, *Quick start guide to industry 4.0: one-stop reference guide for industry 4.0* (CreateSpace Independent Publishing, 2018)
32. S. Misra, *Introduction to industrial internet of things and industry 4.0* (CRC Press, Boca Raton, FL, 2020)

Chapter 6
Industrial IOT in Manufacturing

The real danger is not that computers will begin to think like men, but that men will begin to think like computers.
–Sydney J. Harris.

6.1 Introduction

Internet of Things (IoT) is a network of connected devices like intelligent computers, devices, and objects. It may also be regarded as a worldwide network that connects devices to the Internet and to each other using wired or wireless technology. IoT is expanding rapidly, and it has been estimated that 50 billion devices will be connected to the Internet by 2020. These include smartphones, tablets, desktop computers, autonomous vehicles, refrigerators, toasters, thermostats, cameras, pet monitors, alarm systems, home appliances, insulin pumps, industrial machines, intelligent wheelchairs, wireless sensors, mobile robots, etc. [1]. The growth of Internet of Things (IoT) is drastically making impact on home and industry. The IoT has amalgamated hardware and software to the Internet, thereby creating a smarter world. The application of IoT to the manufacturing industry is known as industrial Internet of Things (IIoT).

While the IoT affects transportation, manufacturing healthcare, or smart homes, the industrial Internet of Things (IIoT) refers to particular applications of IoT to industrial environments. IIoT is a new industrial ecosystem that combines intelligent and autonomous machines, advanced predictive analytics, and machine-human collaboration to improve productivity, efficiency, and reliability. It is bringing about a world where smart, connected embedded systems and products operate as part of larger systems. Like IoT, the Industrial IoT covers many industries and applications. It opens plenty of opportunities in automation, optimization, manufacturing, agriculture, mining, transportation and intelligent vehicle highway systems (IVHS), artificial intelligent (AI) industry, chemical industry, nuclear industry, agricultural industry, mining industry, logistics, oil and gas, transportation, energy/utilities, chemical, aviation, aerospace, and other industrial sectors.

© The Author(s), under exclusive license to Springer Nature Switzerland AG 2023
M. N. O. Sadiku et al., *Emerging Technologies in Manufacturing*,
https://doi.org/10.1007/978-3-031-23156-8_6

Manufacturing is the largest IIoT market. It is also the largest industry from an IoT spending (software, hardware, connectivity, and services) perspective. Manufacturing is among the industrial sectors that will be directly impacted by the disruption springing from IIoT. A smart production unit may consist of a large connected industrial system of materials, parts, machines, tools, inventory, and logistics that can relay data and communicate with each other [2].

This chapter provides application of IIoT in manufacturing. It begins by providing an overview on IoT and IIoT. It presents some applications of IIOT in manufacturing. It highlights the benefits and challenges of applying IIoT in manufacturing. It covers global adoption of IIoT in manufacturing. The last section concludes with comments.

6.2   Overview of Internet of Things

The Internet of Things (IoT) is a new paradigm that promises to allow a variety of things (such as cars, refrigerators, microwaves, thermostats, mobile devices, machines, animals, people, etc.) to be augmented with networking and sensing capabilities, enabling them to work together. It is a global connected network that allows people and devices to interact. The term, Internet of Things, was first coined by Kevin Ashton, a British entrepreneur in 1999. He meant to represent the concept of computers and machines with sensors, which are connected to the Internet to report status and accept control commands [3].

IoT (also known as sensor network or industrial Internet) is a global network of interconnected devices (such as sensors, actuators, personal electronic devices, laptops, tablets, digital cameras, smart phones, alarm systems, home appliances, or industrial machines, and other smart devices) that are enabled with technology of interacting and communicating with each other. It mainly enables the interconnection of Thing to Thing (T2T), Human to Thing (H2T), and Human to Human (H2H). By collecting and combining data from various IoT devices and using big data analytics, decision-makers can take appropriate actions with important economic, social, and environmental implications.

The IoT can be divided into three layers [4]: perception (or sensing) layer, network layer, and application layer.

1. The perception layer collects pertinent and sensorizable information from devices, RFID tags and readers, camera, GPS, and sensors. In this layer, the wireless smart systems with sensors can automatically sense and exchange information among different devices and remotely control them. Any contrivance can have sensors attached to them: people, machines, vehicles, robots, production line, etc.
2. The network layer is mainly messaging and processing information. The role of this layer is to connect all Internet of Things (IoT) compatible devices together and allow them to share the information with each other.

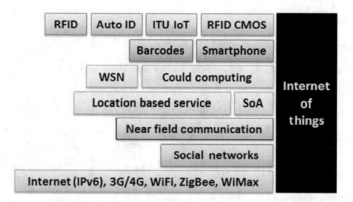

Fig. 6.1 Technologies associated with IoT [5]

3. The application layer is the Internet of Things and the application systems.

Technologies associated with IoT are shown in Fig. 6.1 [5]. These include radio frequency identification (RFID), wireless sensor networks (WSN), middleware, cloud computing, and IoT application software. These five IoT technologies are widely used for the deployment of successful IoT-based products and services. In addition to these, software-defined networking (SDN) is a key enabling technology of industrial Internet of Things [6].

IoT is being implemented in healthcare, wearables, smart cities, smart home, agriculture, automotive, aviation, aerospace, textile, transportation, and manufacturing industries. It has become a powerful force for manufacturing companies. Harnessing the IoT in the world of manufacturing leads to limitless possibilities. Many manufacturers use IoT in conjunction with sensors and employ IoT devices to improve efficiencies and safety. IoT technologies are designed to facilitate and optimize manufacturing processes. They are expanding from household and business use cases to industrial applications. The issue of growing competition and inexpensive connectivity has made IoT crucial and relevant in manufacturing.

6.3   Industrial Internet of Things

The industrial Internet of Things (IIoT) is basically an application of IoT across several industries, such as manufacturing (Industry 4.0), logistics, oil and gas, transportation, energy, mining, aerospace, aviation, automotive, textile, and other industrial sectors. It can be regarded as machines, computers, and people enabling intelligent industrial operations using advanced data analytics. It is a network of systems, objects, platforms, and applications that can communicate and share intelligence. It may also be regarded as a network of intelligent devices connected to form systems that monitor, collect, exchange, and analyze data. It is the biggest and most important part of the overall Internet of Things picture.

Fig. 6.2 IIoT is the integration of Industry 4.0 and the Internet of Things [7]

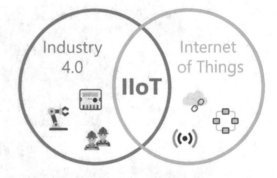

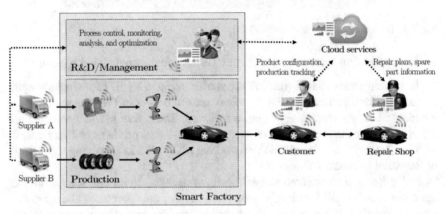

Fig. 6.3 A typical industrial Internet of Things [10]

IIoT is often used in the context of Industry 4.0, the industrial Internet, and related initiatives across the globe. Figure 6.2 shows that IIoT is the integration of Industry 4.0 and the Internet of Things [7]. Industry 4.0 describes a new industrial revolution with a focus on automation, innovation, data, cyber-physical systems, processes, and people [8]. With Industry 4.0, the fourth industrial revolution is set on merging automation and information domains into the industrial Internet of Things, services, and people. The communication infrastructure of Industry 4.0 allows devices to be accessible in barrier-free manner in the industrial Internet of Things, without sacrificing the integrity of safety and security [9]. A typical industrial Internet of Things is shown in Fig. 6.3 [10].

The term "industrial Internet" was coined by Industrial giant GE to describe industrial transformation in the connected context of machines, cyber-physical systems, advanced analytics, AI, people, cloud, and so on. GE and the Industrial Internet Consortium (IIC) decided that IIoT was a synonym for the industrial Internet. IIoT is poised to bring unprecedented opportunities to business and society. Organizations like IIC and IEEE are working concertedly and synergistically to define and develop the IIoT.

IIoT can be implemented on industrial equipment, personnel, and processes, which are all interconnected. IIoT connectivity drives the convergence of operational technology (robots, conveyor belt, smart meters, generator, etc.) and information technology. Within manufacturing, intelligent sensors, distributed control, and secure software are the glue. Forward-thinking manufacturers connect their products to IIoT. They will position themselves as future leaders, while those that fail to act now risk being left behind [11]. Some innovation is necessary to improve the quality and make industrial IoT applications cost-effective.

6.4   Applications

The industrial Internet of Things, a subset of the Internet of Things, is expected to transform many industries including manufacturing. Fig. 6.4 illustrates some of the applications of industrial Internet of Things [12]. IIoT is having a profound effect on the manufacturing sector, leading to the following applications [13, 14]:

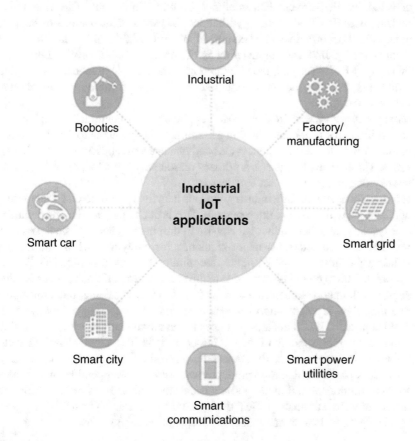

Fig. 6.4   Applications of industrial Internet of Things [12]

- *Automation:* The IIoT is an enabler of industrial automation. Traditionally, plant networks are isolated from each other and from business networks. Now, we can use IoT to connect everything (people, devices, equipment, etc.) within the factory and provide connectivity, communication, and information sharing. This will help manufacturers to automate the production process, thereby eliminating human intervention.
- *Supply Chain Management*: IoT devices can track and trace the inventory system on a global scale in real-time. The IIoT-enabled systems can be configured for location tracking and remote monitoring of inventory. Manufacturers can access real-time supply chain information by using IIoT to track materials and products as they move through the supply chain. IIoT will reduce the expenditure due to mismanagement in the organization. Smart supply chain management solutions using IIoT can enable manufacturers to have real-time insights into the location, status, and condition of every object and process.
- *Smart Automotive Manufacturing:* The smart manufacturing enterprise consists of smart machines, plants, and operations all with intelligence. As the global market compels manufacturers to reconsider operations, smart manufacturing powered by IIoT-driven data analytics becomes important. The automotive industry uses IIoT devices along with industrial robots in the manufacturing process. Robots are effectively and commendably reducing downtime in automotive manufacturing. IIoT can automate many of complex process involved in manufacturing. Mobile sensors, cloud computing, and new applications are helping industrial IoT to become essential and indispensable to automotive manufacturing.
- *Manufacturing Plants*: IIoT has many applications in manufacturing plants. It can facilitate the production flow in a manufacturing plant, as IoT devices automatically monitor development cycles and manage warehouses as well as inventories. The IIoT deployment is a primary reason why investment in IoT devices has skyrocketed over the past few decades.
- *Mining:* Today, the mining industry depends heavily on commercial systems and applications which are not interoperable. This is due to the fact that the commercial systems have their individual non-IIoT compliant technology stack and data formats. This constitutes a major challenge for applying IIoT in the mining industry and slows production. Thus, the mining industry has unique challenges in comparison to other industries due to the infrastructural limitations at the mine sites, which may be underground or surface. In modern mining, there must be a real-time flow of information between enterprise level and shop floor systems. The adoption of IIoT standards practices in the mining industry is an uphill task due to the complex nature of the mining operations. The adoption of IIoT technology in the mining industry offers safer mine site for workers, makes mining operations predictable, boosts productivity, provides interoperable environment for both traditional and modern systems/devices, and reduces human error and intervention by automating. IIoT deployment in the mining industry improves efficiency, decreases operational costs, and reduces energy usage.

Other applications of IIoT in manufacturing include digital supply chain, retail, automotive, unmanned aerial vehicles or drones, aerospace, aviation, agriculture, predictive maintenance, data-enabled services, connected logistics, smart homes, smart grid, smart city, smart farming, energy consumption optimization, safety, and health monitoring of workers.

6.5 Benefits

Industrial IoT will have huge and beneficial impacts in manufacturing. It can enhance productivity, boost operational performance, increase efficiency, increase throughput, assist asset monitoring, minimize financial risk, reduce production downtime or boost the production speed, improve product quality, reduce overhead, conserve resources, increase profits, eliminate waste, support better customer experience, and improve employee health and safety. Other benefits include [15]:

- *Enhancing Productivity:* All the investment on building connectivity may seem futile if one cannot make sense of the data collected, therefore emphasizing the great importance of IIoT deployment for productivity improvement. Monitoring the whole manufacturing process digitally and economically helps in boosting the productivity of the manufacturing industry.
- *Workplace Safety:* This is a top concern in the manufacturing industry. An advantage of IIoT is its ability to increase safety in manufacturing. A workers' safety in the manufacturing environment can improve by deployment of IoT combined with big data analytics. IIoT system can help assure employees' safety and equipment uptimes and equipment utilization by monitoring equipment for potentially dangerous failures.
- *Well-Being of Workers:* Proactive manufacturers see the pandemic-related investments as a great platform to offer safety of coworkers by tracking their location. Wearable ear devices that can allow workers to set their own soundtrack can boost employee safety and worker on-the-job satisfaction ratings.
- *Predictive Maintenance:* This will result from IoT integration. Manufacturing companies can use real-time data generated from IIoT systems to predict when an equipment will need service. IIoT takes a preventive maintenance approach to the next level by saving manufacturers a lot of money. By using sensors, cameras, and data analytics, managers can determine when a piece of machinery will fail before it actually does. Figure 6.5 illustrates predictive maintenance with IoT [16].
- *Customer Satisfaction:* The issue of the customer value always comes before selecting technical solutions. IIoT technology is having a great impact on customer experience by enhancing customer satisfaction. It is ushering in a new age of opportunity, enabling businesses to drive customer relationships that are closer and more profitable than ever before.

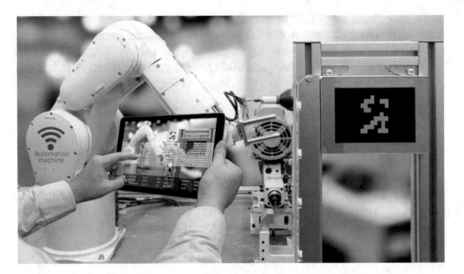

Fig. 6.5 Predictive maintenance with IoT [16]

- *Asset Management*: IIoT solves some of the challenges facing asset-intensive companies. An automotive manufacturer can take advantage of IIoT to track its assets, including automotive fleet, tools, and vehicle parts. IIoT can also track asset locations with smart sensors, monitor demand-supply requirements, and manage the production accordingly. IIoT is applied in manufacturing to ensure proper asset usage and provide the best return on assets.
- *Quality Control and Long Run Organizational Sustainability:* IoT sensors collect data from various stages of a product life cycle. The IoT device can provide data about the customer satisfaction levels, sentiments, customer experiences, and suggestions after using the product. These inputs can be analyzed to identify and correct quality, reliability, and safety issues, which can lead to significant improvement, long-term customer satisfaction, and loyalty resulting in the long run organizational sustainability

Some of these benefits are depicted in Fig. 6.6 [17]. Other benefits include cost reduction, shorter time-to-market, energy management, and mass customization. Manufacturing organizations that intend to modernize cannot ignore the benefits of IoT and IIoT technology deployments.

6.6   Challenges

The industrial Internet of Things is still in its early age. Although it offers several benefits for manufacturing and is poised to grow significantly, there are challenges which can hinder future growth. Some of the challenges facing the adoption of IIoT in manufacturing include [18]:

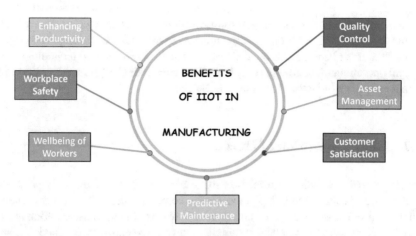

Fig. 6.6 Some benefits of IIoT in manufacturing. (Modified [19])

- *Security:* A major challenge for executives is cybersecurity and data security, which is rising in importance due to increased vulnerability to attacks and data breaches. Connecting industrial machines to an interconnected system is risky. A secure environment is necessary to protect machines and employees from cyber threats. In the IIoT context, data is considered sensitive because data will encapsulate various aspects of industrial operation, including highly sensitive information about products, business strategies, and companies. The transition to more open network architectures and data sharing of IoT poses challenges in industrial markets. The loss of sensitive information can lead to significant business loss and cause reputational damage [19].
- *Interoperability:* The manufacturing processes deployed in different industries vary based on such factors as industry positioning and prevailing industrial standards, customers served and customer requirements, etc. Interoperability between sensors, edge devices, and the cloud has always been an important challenge. Standards are not in place to ensure the transferring of data between machines from different vendors.
- *Lack of Standards:* Standards are needed to enable smart, connected machines, products, and assets to interact in a transparent, effective, and synergistic manner. They are vital to ensure that any new device added to the infrastructure can interact seamlessly with existing equipment. The question of standardization is currently being addressed by the Industry 4.0 and the Industrial Internet Consortium [20].
- *Lack of Skills:* A major challenge making companies not to be ready for the industrial Internet of Things is a lack of skilled workers in the IIoT high technology arena. As IIoT is a new area, new industry skills are required and plans need to be implemented to grow the talent corp needed to sustain this new and emerging high technological industrial communications giant.
- *Integration:* Data integration is a major challenge because it is not easy to move from data to business value. The integration of IIoT solutions into existing busi-

ness processes and industrial environment brings new challenges. It is only when the data from IoT solutions is fully integrated with data from enterprise systems that the most benefits can be achieved. Integrating information technology (IT) and operational technology (OT) is another critical challenge faced by the companies during industrial IoT implementation.

6.7   Global Adoption of IIOT

Industrial IoT can radically transform manufacturing. Manufacturing is poised to keep the first position across the globe when it comes to using IoT. IoT deployments offer rapid return and enable manufacturers to realize digital transformations from several perspectives: efficiency, automation, customer-centricity, productivity, and competitive advantage with long-run implications for organizational profitability, growth, and sustainability. We will consider how some nations have implemented IIoT in their manufacturing sector.

- *United States*: Industrial giant GE coined the term industrial Internet which really describes industrial transformation. In that regard, a linguistic parallel can be made with the word coinage, "Internet of Everything," a term Cisco coined and used up until 2016. In many regards, the Internet of Everything, in an industrial context, is closer to industrial IoT. Companies such as GE, IBM, and Cisco operate across the globe and are often members of several industry bodies at the same time.
- *United Kingdom:* The United Kingdom is going through a productivity crisis. For manufacturing companies across the United Kingdom, the benefits of increased operational performance and improved reliability realized through IIoT solutions are compelling. While the United Kingdom's problem with productivity is complex, innovation in technology undeniably has an important role to play in addressing the challenge. IIoT technology provides, which reciprocally holds the potential to have a transformative effect for the manufacturing sector. Through greater efficiency, improved business insights, and better-informed decisions, IIoT technology can help the United Kingdom boost its productivity significantly [21].
- *India*: This is a nation that lacks in adapting to newer technologies. Traditional businesses in India are being disrupted with the advent of newly emerging technologies. In 2020, India contributed just $10 billion out the global IoT market of $373 billion. India is regarded as a hub for human resource. This should propel India to new heights in the field of IIoT [22].
- *China:* Chinese tech giants such as Tencent, Alibaba, JD, and Baidu have all put in place aggressive plans to adopt IIoT. That is merely a tip of the iceberg of China's aggressive IIoT development plan. It is expected that investment drives to reduce China's dependence on foreign technology would continue to be an important trend to watch [23].

6.8 Conclusion

The Internet of Things (IoT) and its subset, the industrial Internet of Things (IIoT), are transforming modern manufacturing. The industrial Internet of Things refers to a vast number of interconnected industrial systems that are communicating, sharing data, and improving industrial performance to benefit the society. Recently, IIoT has emerged as a subparadigm which focuses more on safety-critical applications in industries like aerospace, aviation, automotive, agriculture, food processing, energy, and healthcare. It may be regarded as an important element of the fourth industrial revolution that is changing the face of industry in a profound manner. In manufacturing, IIOT improves productivity, enhances efficiency, and drives competitive advantage.

Companies that want to stay competitive should embrace the IIoT as soon as possible. In virtually every industry, areas can be identified to get started, where implementation costs are low and payoffs are high. More information about IIoT can be found in the books in [22–27] and in the following related journals: *IoT* and *IEEE Internet of Things Journal*.

References

1. M.N.O. Sadiku, *Emerging Internet-Based Technologies* (CRC Press, Boca Raton, 2019)
2. M.N.O. Sadiku, Y. Wang, S. Cui, S.M. Musa, Industrial internet of things. Int. J. Adv. Sci. Res. Eng. **3**(11), 1–4 (2017)
3. M.N.O. Sadiku, S.M. Musa, S.R. Nelatury, Internet of things: An introduction. Int. J. Eng. Res. Adv. Technol. **2**(3), 39–43 (2016)
4. G. Wiggins, Chemistry on the internet: The library on your computer. J. Chem. Inf. Comput. Sci. **38**, 956–965 (1998)
5. L.D. Xu, W. He, S. Li, Internet of things in industries: A survey. IEEE Trans. Ind. Inf. **10**(4), 2233–2243 (2014)
6. S. Al-Rubaye et al., Industrial internet of thing driven by SDN platform for smart grid resiliency. IEEE Internet Things J. (2017)
7. 7 Uses, applications & benefits of industrial IoT in manufacturing (n.d.), https://infinite-uptime.com/blog/industrial-iot-in-manufacturing/
8. Predictive maintenance with IoT: The road to real returns (n.d.), https://www.avnet.com/wps/portal/abacus/solutions/markets/industrial/predictive-maintenance-iot/
9. The industrial Internet of things (IIoT): the business guide to Industrial IoT (n.d.), https://www.i--scoop.eu/internet-of-things-guide/industrial-internet-things-iiot-saving-costs-innovation/
10. D. Schulz, FDI and the industrial Internet of things, in Proceedings of IEEE 20th Conference on Emerging Technologies & Factory Automation, 2015, pp. 1–8
11. A.R. Sadeghil, C. Wachsmann, and M. Waidner, Security and privacy challenges in industrial Internet of things, in Proceedings of the 52nd Annual Design Automation Conference, 2015
12. The industrial Internet of things: Why it demands not only new technology – But also a new operational blueprint for your business (n.d.), https://www.pwchk.com/en/migration/pdf/tmt-industrial-internet-may2016.pdf
13. Industrial Internet of things (IIoT) (n.d.), https://internetofthingsagenda.techtarget.com/definition/Industrial-Internet-of-Things-IIoT

14. S. Morgan, Top 3 Applications of industrial IoT in manufacturing (n.d.), https://stumpblog.com/top-3-applications-of-industrial-iot-in-manufacturing/
15. A. Aziz, O. Schelén, U. Bodin, A study on industrial IoT for the mining industry: Synthesized architecture and open research directions. IoT **1**, 529–550 (2020)
16. How industrial IoT (IIoT) is revolutionizing manufacturing industry (n.d.), https://iot.do/how-industrial-iot-iiot-is-revolutionizing-manufacturing-industry-2021-01
17. K.S. Wong, M.H. Kim, Privacy protection for data-driven smart manufacturing systems. Int. J. Web Serv. Res. **14**(3), 17–32 (2017)
18. J. Conway, The industrial Internet of things: An evolution to a smart manufacturing enterprise (n.d.), http://www.mcrockcapital.com/uploads/1/0/9/6/10961847/schneider-an_evolution_to_a_smart_manufacturing_enterprise.pdf
19. How to succeed with industrial IoT in manufacturing (n.d.), https://business-reporter.co.uk/2019/07/15/how-to-succeed-with-industrial-iot-in-manufacturing/
20. What is Industrial IoT or IIoT? (n.d.), Unknown Source
21. IIoT: Enterprise digital transformation and Industry 4.0 in China (2020), https://www.premia-partners.com/insight/iiot-enterprise-digital-transformation-and-industry-40-in-china
22. G. Veneri, A. Capasso, *Hands-on Industrial Internet of Things: Create a Powerful Industrial IoT Infrastructure Using Industry 4.0* (Packt Publishing, 2018)
23. S. Bhattacharjee, *Practical Industrial Internet of Things Security: A Practitioner's Guide to Securing Connected Industries* (Packt Publishing, 2018)
24. S. Misra, *Introduction to Industrial Internet of Things and Industry 4.0* (CRC Press, Boca Raton, 2020)
25. S. Jeschke et al., *Industrial Internet of Things: Cybermanufacturing Systems* (Springer, 2016)
26. A. Gilchrist, *Industry 4.0: The Industrial Internet of Things* (Apress, 2016)
27. S. Goundar et al., *Innovations in the Industrial Internet of Things (IIoT) and Smart Factory* (IGI Global, 2021)

Chapter 7
Big Data in Manufacturing

Every company has big data in its future and every company will eventually be in the data business.

–Thomas H. Davenport

7.1   Introduction

The manufacturing industry adds a lot of value to any nation's economy. Manufacturing is critical to national economies by providing jobs and improving the quality of life through innovation. It is well-known that manufacturing represents the most challenging and complex industry as far as the variety of its products is concerned. Most industrial manufacturing companies have complex manufacturing processes often with equally complex supply chain and suppliers. There are different types of manufacturing including biopharmaceutical manufacturing, chemical manufacturing, discrete manufacturing, aerospace manufacturing, advanced manufacturing, smart manufacturing, cloud manufacturing, offshore manufacturing, lean manufacturing, green manufacturing, sustainable manufacturing, distributed manufacturing, predictive manufacturing, computer-integrated manufacturing, and computer-aided manufacturing.

Manufacturers of all types of products are finding significant value in big data. In the data-driven economy, turning data into actionable analytics is the best way to boost efficiency, safety, quality, and productivity. Modern manufacturing facilities are data-rich environments that exploit and thrive on manufacturing intelligence. The potential benefits of manufacturing intelligence include improvements in operational efficiency, process innovation, and environmental impact [1].

Big data refers to the large volume of structured and unstructured data with the implicit potential to be mined for information. Big data is everywhere, and the big data revolution is upon us. Data has been helping businesses make better decisions for centuries. As shown in Fig. 7.1, businesses from all types of industries have greatly benefited by adopting big data solutions [2]. This is true of healthcare, energy, finance, telecommunication, marketing, and sports. The manufacturing industry has joined the big data bandwagon as well. The application of big data technologies in the manufacturing sector is a relatively new interdisciplinary

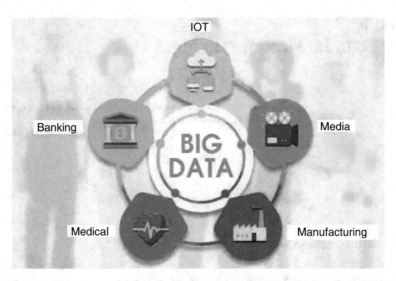

Fig. 7.1 Real-time applications of big data [2]

research area which incorporates automation, engineering, and data analytics [3]. Big data has been a fast-changing research area with many new opportunities for applications in manufacturing. Big data analytics can provide the manufacturing industry with the imperative to succeed in an increasingly complex environment.

This chapter provides an introduction to the use of big data in manufacturing. It begins by giving an overview on the characteristics of big data. It covers big data analytics. It describes the role of big data in manufacturing. It presents some applications of big data in manufacturing. It highlights the benefits and challenges of big data in manufacturing.

It presents global adoption of big data in manufacturing. The last section concludes with comments.

7.2 Big Data Characteristics

Big data (BD) is the umbrella term used to define large chunks of data found in industrial machines, government filings, medical records, and other sources. It includes complex and diverse streams of data obtained from various sources and requires advanced processing techniques. It is a relatively newer technology that can help the manufacturing industry. The three main sources of big data are machines, people, and companies. Big data can be described with 42 Vs [4]. The first five Vs are volume, velocity, variety, veracity, and value [5].

- *Volume*: This refers to the size of the data being generated both inside and outside organizations and is increasing annually. Some regard big data as data over one petabyte in volume.
- *Velocity*: This depicts the unprecedented speed at which data are generated by Internet users, mobile users, social media, etc. Data are generated and processed in a fast way to extract useful, relevant information. Big data could be analyzed in real time, and it has movement and velocity.
- *Variety*: This refers to the data types, since big data may originate from heterogeneous sources and is in different formats (e.g., videos, images, audio, text, logs). BD comprises of structured, semi-structured, or unstructured data.
- *Veracity*: By this, we mean the truthfulness of data, i.e., whether the data comes from a reputable, trustworthy, authentic, and accountable source. It suggests the inconsistency in the quality of different sources of big data. The data may not be 100% correct.
- *Value*: This is the most important aspect of the big data. It is the desired outcome of big data processing. It refers to the process of discovering hidden values from large datasets. It denotes the value derived from the analysis of the existing data. If one cannot extract some business value from the data, there is no use managing and storing it.

On this basis, small data can be regarded as having low volume, low velocity, low variety, low veracity, and low value. Additional five Vs have been added [6]:

- *Validity:* This refers to the accuracy and correctness of data. It also indicates how up-to-date it is.
- *Viability:* This identifies the relevancy of data for each use case. Relevancy of data is required to maintain the desired and accurate outcome through analytical and predictive measures.
- *Volatility:* Since data are generated and change at a rapid rate, volatility determines how quickly data change.
- *Vulnerability:* The vulnerability of data is essential because privacy and security are of utmost importance for personal data.
- *Visualization:* Data needs to be presented unambiguously and attractively to the user. Proper visualization of large and complex clinical reports helps in finding valuable insights.

Figure 7.2 shows the 10 Vs of big data. Instead of the 10 Vs above, some suggest the following 5 Vs: venue, variability, vocabulary, vagueness, and validity) [7].

To thrive in today's complex business environment, businesses must adopt a data-driven culture and leverage analytics platforms to make key decisions that improve productivity. Industries that benefit from big data include the healthcare, financial, airline, travel, restaurants, automobile, sports, agriculture, manufacturing, and hospitality industries. Big data technologies are playing an essential role in farming: machines are equipped with sensors that measure data in their environment.

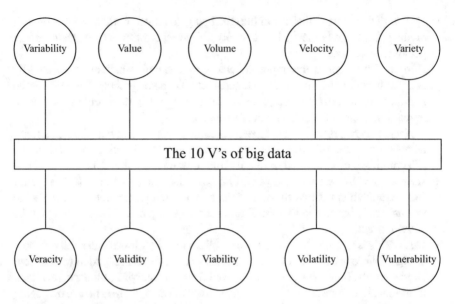

Fig. 7.2 The 10 Vs of big data

7.3   Big Data Analytics

Every day, data is growing bigger and bigger, and big data analysis (BDA) has become a requirement for gaining invaluable insights into data such that companies could gain significant profits in the global market. Once the big data is ready for analysis, we use advanced software programs such as Hadoop, MapReduce, MongoDB, and NoSQL databases [8]. Big data analytics refers to how we can extract, validate, translate, and utilize big data as a new currency of information transactions. It is an emerging field that is aimed at creating empirical predictions. Data-driven organizations use analytics to guide decisions at all levels [9].

Data scientists know how to use tools that identify patterns and relationships that may otherwise remain hidden. They are part of virtually every major industry, and manufacturing is no exception. Manufacturing big data analytics is expected to ensure better farming practices and decision-making and a sustainable future for humankind. This will involve artificial intelligence and machine-learning technologies to determine better manufacturing practices and decision-making [10].

Implementing a big data analytics solution can help almost every part of a manufacturing enterprise. The manufacturer industry is beginning to deploy big data analytics for two essential tasks: gain control of the vast amounts of data generated and use the right information to drive productivity and enhance decision-making [11]. One way manufacturers can take more control of their quality parameters is through data analytics. To stay competitive and keep pace with consumers' fickle buying habits, the manufacturing industry must consider implementing data analytics tools.

A very important attribute of data analytics is that it can be implemented anywhere in the world.

7.4 Big Data in Manufacturing

Manufacturing can be described as a 5 M system which consists of [12]:

- Materials (properties and functions)
- Machines (precision and capabilities)
- Methods (efficiency and productivity)
- Measurements (sensing and improvement)
- Modeling (prediction, optimization, and prevention)

Data has long been the lifeblood of manufacturing. Manufacturing data is key to conducting data-driven manufacturing. Most manufacturing data is standardized which is supported by various industrial vendors and associations. Big data in manufacturing is generated from the production and process, from machines such as pumps, motors, compressors, meters, sensors, controllers, or conveyers. The big data in manufacturing does not mean anything without analysis. In order to benefit and gain insight from the data collected, it must be analyzed and visualized. The big data solutions analyze, collect, and monitor massive amount of unstructured and structured data generated from several sources such as product quality, production unit, and factory floor. When big data is analyzed, it opens a window of opportunity for manufacturers to identify and fix problems before they get worse. Fig. 7.3 shows big data in manufacturing [13].

Fig. 7.3 Big data in manufacturing [13]

Fig. 7.4 Benefits of big data analytics to companies [16]

The big data revolution is rapidly changing the manufacturing industry. Six key drivers of big data applications in manufacturing have been identified as system integration, data, prediction, sustainability, resource sharing, and hardware. Based on the requirements of manufacturing, nine essential components of big data ecosystem have been identified as ingestion, storage, computing, analytics, visualization, management, workflow, infrastructure, and security [14, 15]. In the manufacturing sector, big data can combine and analyze real-time information in every step of the production process. The manufacturing industry adopts big data technology to design, build, and distribute products. Typical examples of manufacturers using big data analytics are oil and gas companies, refineries, chemical producers, and automotive industries. Fig. 7.4 shows some benefits of big data analytics to companies [16].

7.5   Applications

The data in manufacturing is often used for operational purposes manufacturing. Fig. 7.5 shows how car manufacturers use big data [17]. Different applications of big data in manufacturing include [18]:

- *Preventive Maintenance:* Operational efficiency relies on the availability of the machinery in the production process. Manufacturers can use the best big data analytics software to determine the status of the machine. This approach is often called preventive maintenance. Most manufacturers often follow some schedule of preventative maintenance. Being able to predict equipment failure can save manufacturers significant maintenance time and boost production rates.
- *Product Design:* Developing new products can be expensive. Big data analytics software can analyze data and help identify trends and market changes that can be fed into the design of new products.

Fig. 7.5 How car manufacturers are using big data [17]

- *Production Management Automation*: Automation of the production management is probably the hardest way of using big data in manufacturing processes. A good example of production management automation is the case with *GE's wind turbines*. Sensors provide data on energy generation and wind direction, according to which the blade pitch is changed to optimize the wind turbine's efficiency.
- *Customer Experience:* Excellent customer experience has become an essential part of every business. The big data analytics solutions draw customer data from a wide variety of sources and use the data for in-depth data analysis. Manufacturers now use big data analytics solutions to examine social media, customer service, and marketing data.
- *Supply Chain Improvement:* Timing is everything. Big data enables manufacturers to track the exact location of their products. Modern supply chains are becoming increasingly complex. Big data analytics solutions deliver supply chain visibility to instantly know key supply chain information. Big data can provide information about the supply chain and makes it possible to predict with greater certainty whether or not a supplier will deliver as agreed. It also increases transparency into the entire supply chain.
- *Quality Assessment:* Quality is a priority to manufacturers. Maintenance of product quality is essential to manufacturers. Quality control is an area that has traditionally remained in the realm of human operations. With customers demanding more customization, quality control assessment has become important for manufacturing. With the advent of big data analytics, automatic quality testing is helping to drive production efficiencies, improve production quality, save time, and help avoid human errors. Many manufacturers have turned to big data to help them improve their quality assurance. By using big data and advanced

analytics, manufacturing companies can view product quality and delivery accuracy in real-time. Big data analytics can help manufacturers keep up with quality metrics and standards. Big data can also be used to reduce wastage. Using big data analytics can help managers identify whether some products are good or not. Using big data and advanced analytics, manufacturers are able to view product quality and delivery accuracy in real-time.

The range of applications of big data in the manufacturing industry is limited only by the creativity and imagination of the big data analyst, technology leaders, and business executives.

7.6   Benefits

With big data analytics, a company can improve manufacturing, customize product design, ensure better quality assurance, manage the supply chain, and evaluate for any potential risk [19]. With many manufacturers setting the goal of getting the right quantity of products to the right market at the right time and at the right price, they need to consider harnessing big data to gain actionable insights, make better decisions, and boost the bottom line. Big data analytics can help businesses to make informed and timely decisions, which are crucial for business improvement and growth. In a very competitive environment, manufacturers can turn to big data to improve their business operations. Big data can take a manufacturer from predictive decision-making to prescriptive decision-making, thereby leveraging significant competitive advantages with implications for increased revenues, profitability, and business growth.

Other benefits include [15, 18, 20, 21]:

- *Lifeblood of Manufacturing*: Big data can help gain insights into your manufacturing operations and reveal the glitches in the business operations. Its analysis opens a window of opportunity for manufacturers to identify and fix problems.
- *Reduced Process Flaws:* Data analytics can greatly improve assembly-line efficiency, stream process flows, and support corrective and preventive maintenance operations thereby further reducing process flaws.
- *Speeding Up Assembly*: Big data helps manufacturing companies to know where they are most efficient, with the added possibility of producing more products in that area.
- *Improved Workforce Efficiency:* Big data in manufacturing can improve productivity, growth, and workforce efficiency.
- *Competitive Advantage:* It is becoming a crucial way for companies to outperform their peers by leveraging data-driven strategies.
- *Increased Efficiency:* Manufacturing big data can greatly improve production line speed and quality, which helps a manufacturing company control costs, increase productivity, and boost margins.
- *Improved Customer Service*: Big data analytics solutions can help manufacturers better gauge customer sentiment and respond to customers in real time.

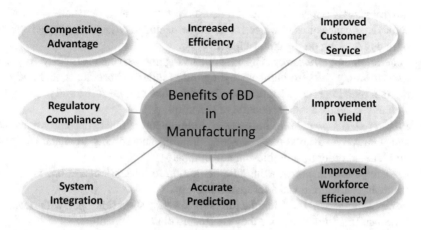

Fig. 7.6 Benefits of big data in manufacturing

- *Regulatory Compliance:* Data analytics can be effectively used to ensure regulatory compliance.
- *Improvements in Yield:* Big data analytics can be a critical tool for realizing improvements in yield, particularly in a manufacturing environment with process complexity, process variability, and capacity restraints. Big data can improve product accuracy during the production process.
- *System Integration:* Using big data technologies for system integration is a crucial enabler of smart manufacturing. This allows for integrating and cooperating manufacturing systems to timely adapt dynamic demands from production and supply chain.
- *Accurate Prediction:* This enables manufacturing to change from the reactionary to the preventive mode of operations. Due to the increasing use of big data analytic in manufacturing, it is feasible to predict the behaviors of individual or linked manufacturing systems accurately. Predictive modeling helps in better forecasting of a host of business aspects.

Figure 7.6 illustrates some of these benefits.

7.7 Challenges

As far as big data is concerned, the manufacturing industry has its share of challenges to contend with. Similar to big data, manufacturing organizations face the same challenges associated with 5 Vs of big data (volume, velocity, variety, veracity, and value). The challenges include the following [20, 22, 23]:

- *Complexity:* Due to the complex nature of manufacturing, adoption of latest technologies is always difficult.

- *Environmental Issues*: As suits the occasion, the incident, or where warranted, some drastic measures may need to be implemented to reduce the adverse effects of manufacturing on the environment.
- *Regulation Requirements:* The manufacturing industry inadvertently exposes itself to safety risks (to humans) arising out of industrial accidents, and mitigating this is an imperative for safeguarding the value of human life and property as well as the avoidance or reduction of litigation.
- *Product Development and Innovation:* Manufacturers must constantly innovate and creatively advance news ways of thinking and solving of manufacturing problems in order to stay relevant and compete.
- *Preventive Maintenance*: Minimum disruption to business operations is one of the main challenging tasks for manufacturers.
- *Faster Time-to-Market:* Manufacturers have a very short time span within which to launch their products, else they risk losing out to competitors.
- *Market Volatility:* Market volatility constrains enhanced product performance, thus impeding business growth, sustainability, and profitability.
- *Big or Huge Data and Manual Analysis:* Manual analysis of big or huge data is undesirable as it is inherently time-consuming, labor intensive, error-prone, and risk laden (human errors).
- *Judicious Use:* Big data often leads managers to rely too much on the data and abdicate the very important role of sound and insightful decision-making. They need to be judicious in their use of data especially around privacy and sensitivity concerns.
- *Cybersecurity:* With the increasing adoption of emerging technologies in the manufacturing industry, the security concerns are becoming significant. Manufacturing is one of the most targeted industries by cyberattackers owing to the presence of vital data, making manufacturers subjected to cybersecurity incidents. Perhaps the smartest way that organizations can address the huge demand for cybersecurity skills is to invest in cybersecurity. Cybersecurity is an essential and advanced technology needed to protect data assets in manufacturing. Areas that are vulnerable in both the physical and cyberspace of the manufacturing systems must be identified and protected.
- *Standardization:* Standardization of big data platform improves the feasibility of enterprise-ready solutions in manufacturing. Cybersecurity will continuously challenge manufacturing because security standards are still not available in some systems.
- *Ethical Issues:* There are four major ethical challenges in manufacturing: (1) privacy, sharing of personal information without permission; (2) security, protection of data from outside threats; (3) ownership, the rightful ownership of the data used for analytics; (4) evidence-based decision-making, the use of data to make decisions about a population based solely on quantitative information.

7.8   Global Adoption of Big Data in Manufacturing

The manufacturing sector has been the backbone of many developed and developing markets. Using big data analytics in manufacturing, companies can tackle global development challenges, such as strategically transferring production to other countries or opening new factories in new and potentially very profitable locations. In developed economies, manufacturers can use big data to reduce costs and deliver greater creativity and innovation in products and services. Industry classification benchmark organizations such as the North American Industry Classification System (NAICS) and International Standard of Industrial Classification (ISIC) have categorized the manufacturing industry into two major segments based on the production processes they follow—discrete manufacturing and process manufacturing. We now consider how some nations adopt big data in their manufacturing sector.

- *United States:* Manufacturing sector adds great value to the US economy. The United States is among the lead innovators and users of big data analytics in the manufacturing industry and is expected to hold a significant share over the forecast period. According to the MAPI (Manufacturers Alliance for Productivity and Innovation), the manufacturing sector is forecast to increase faster than the general economy. For example, American multinational Intel corporation has been harnessing big data in support of its processor manufacturing operations for some time. It is using big data to develop chips faster, identify manufacturing glitches, and advance warnings about security threats. Its factory equipment generated data into their big data solution. The analytics solution uses this data for artificial intelligence (AI)-driven pattern recognition, fault detection, and visualization. Their predictive maintenance reduces reaction time from 4 hours to 30 seconds and cuts costs. Intel predicted saving $100 million in 2017 [15, 24].
- *United Kingdom:* The UK defense minister recently confirmed that sponsored cyberattacks on the United Kingdom's infrastructure could cause economic chaos. The United Kingdom used to lead the world in cybersecurity expertise. Today, government representatives are searching for expertise and skills across the globe. Home-grown talent is not being developed enough. Organizations in the United Kingdom now realize the need to invest heavily in security [25].
- *Germany:* Manufacturing is regarded as an important factor for the European economy and also for societal development and growth. Industry 4.0 is a German government initiative that promotes automation of the manufacturing industry with the goal of developing smart factories. Big data is already being used for optimizing production schedules based on supplier, customer, machine availability, and cost constraints.

7.9   Conclusion

The big data era has emerged. Data has long been the lifeblood of manufacturing. In order to become more competitive and improve their efficiency and productivity, manufacturers must embrace big data and big data analytics. Use of data analytics to make decisions about business strategy and operations is becoming common-place. Today, leveraging big data has become a business imperative since it is enabling solutions to long-standing business challenges for manufacturing compa-nies worldwide. The leading manufacturers are poised to use big data technologies to capitalize on this and many other sources and expand their customer base across foreign countries. For those manufacturing businesses that are still wondering what big data can do for them, the sooner they get to using big data analytics, the sooner they will be able to apply the latest innovations in data science in support of their manufacturing operations. Once they do so, the sky's the limit. More information on big data in manufacturing can be found in the books in [26, 27] and related journals: *Journal of Big Data* and *Manufacturing Letters*.

References

1. P. O'Donovan et al., Big data in manufacturing: a systematic mapping study. J Big Data **2**, 20 (2015)
2. Real-time applications of big data that drive various industries, https://elysiumpro.in/applications-big-data-drive-industries/
3. M.N.O. Sadiku, J.T. Ashaolu, A. Ajayi-Majebi, S.M. Musa, Big data in manufacturing. Int J Sci Adv **2**(1), 63–66 (2021)
4. The 42 V's of big data and data science, https://www.kdnuggets.com/2017/04/42-vs-big-data-data-science.html
5. M.N.O. Sadiku, M. Tembely, S.M. Musa, Big data: An introduction for engineers. J Sci Eng Res **3**(2), 106–108 (2016)
6. P.K.D. Pramanik, S. Pal, M. Mukhopadhyay, Healthcare big data: A comprehensive overview, in *Intelligent Systems for Healthcare Management and Delivery*, ed. by N. Bouchemal, (IGI Global, chapter 4, 2019), pp. 72–100
7. J. Moorthy et al., Big data: Prospects and challenges. J Decis Mak **40**(1), 74–96 (2015)
8. M.N.O. Sadiku, J. Foreman, S.M. Musa, Big data analytics: A primer. Int J Technol Manag Res **5**(9), 44–49 (2018)
9. C.M.M. Kotteti, M.N.O. Sadiku, S.M. Musa, Big data analytics. IJRTEM **2**(10), 2455–3689 (2018)
10. M. Ryan, Agricultural big data analytics and the ethics of power. J Agric Environ Ethics **33**, 49–69 (2020)
11. Top 5 big data analytics software benefits for manufacturing in 2019, https://blogs.opentext.com/big-data-analytics-benefits-manufacturing/
12. J. Lee et al., Recent advances and trends in predictive manufacturing systems in big data envi-ronment. Manufac Lett **1**, 38–41 (2013)
13. A. Bekker, Big data in manufacturing: Use cases + Guide on how to start, https://www.scnsoft.com/blog/big-data-in-manufacturing-use-cases
14. K. Nagorny et al., Big data analysis in smart manufacturing: A review. *Int J Commun Netw Syst Sci* **10**, 31–58 (2017)

15. Y. Cui, S. Kara, K.C. Chan, Manufacturing big data ecosystem: A systematic literature review. Robot Comput Integr Manuf **62**, 101861 (2020)
16. T. K. Arachchi, Why big data? May 2020, https://towardsdatascience.com/why-big-data-bf0d65933782
17. R. Delgado, How car manufacturers are using big data, June 2015, https://datafloq.com/read/car-manufacturers-are-using-big-data/1204
18. Top 5 big data analytics software benefits for manufacturing in 2019, January 2019, https://blogs.opentext.com/big-data-analytics-benefits-manufacturing/
19. L. Andrews, What are examples of big data in manufacturing? October 2019, https://www.sensrtrx.com/what-are-examples-of-big-data-in-manufacturing/
20. Challenges of the manufacturing industry, & big data analytics, August 2017, https://www.msrcosmos.com/blog/challenges-of-the-manufacturing-industry-big-data-analytics/
21. R. Delgado, Applications of big data in manufacturing, April 2016, https://www.automation.com/en-us/articles/2016-1/applications-of-big-data-in-manufacturing
22. H.N. Dai et al., Big data analytics for manufacturing Internet of things: Opportunities, challenges and enabling technologies. Enterp Inf Syst **14**(9-10), 1279–1303 (2020)
23. V. Chang, W. Lin, How big data transforms manufacturing industry: A review paper. Int J Strat Eng **2**(1), 39 (2019)
24. Big data analytics in manufacturing industry set to exceed $4.5 Billion by 2025 - Condition monitoring to register significant growth, April 2020, https://www.prnewswire.com/news-releases/big-data-analytics-in-manufacturing-industry-set-to-exceed-4-5-billion-by-2025%2D%2D-condition-monitoring-to-register-significant-growth-301033518.html
25. G. Gupta, Knowing the limitations of big data, May 2020, https://technologymagazine.com/data-and-data-analytics/knowing-limitations-big-data
26. S. Khanna, *Big data in manufacturing: a practical introductory guide to starting your big data journey* (Kindle Edition, 2017)
27. S.M.P. Hosseini, A. Azizi, *Big data approach to firm level innovation in manufacturing: industrial economics* (Springer, 2020)

Chapter 8
Additive Manufacturing

Additive manufacturing could reduce energy use by 50 percent and reduce materials costs by up to 90 percent compared to traditional manufacturing.

–Anonymous

8.1 Introduction

Manufacturing is essentially about converting raw material into products and services.

Today, we are witnessing a manufacturing trend toward miniaturization. As a result, there is an ever-increasing demand for miniaturized components. Additive manufacturing (AM) has emerged as a prototyping method. It is a new technology to automatically produce three-dimensional (3D) objects by printing successive layering of material. It goes by other names such as free form fabrication, rapid prototyping (RP), 3D printing, or direct digital manufacturing. **It** describes the technologies for building 3D objects by adding layer upon layer of material, and the material could be plastic, metal, concrete, or human tissue. AM is often regarded as an Industry 4.0 technology [1].

AM enables cost-efficient and rapid fabrication of complex components from material. The material could be supplied in the form of a powder or wire coupled or spread into a laser or electron beam chamber, melted, and then deposited selectively. As shown in Fig. 8.1, AM is encapsulated into the three Cs: complexity, customization, and consolidation [2].

As its name implies, additive manufacturing adds material to create an object, and it inherently produces less waste. While traditional production systems proceed by subtraction, additive manufacturing involves fabricating a 3D object by successively adding material, usually in a layer-by-layer fashion, as shown in Fig. 8.2 [3]. AM is a process of building an object one thin layer at a time.

This chapter provides a brief introduction to additive manufacturing for readers with both a technical and business mindset. It begins by describing what additive manufacturing is all about. It describes seven additive manufacturing processes. It presents some application of additive manufacturing. It highlights the benefits and challenges of AM. It covers the global adoption of AM. The last section concludes with comments.

© The Author(s), under exclusive license to Springer Nature Switzerland AG 2023
M. N. O. Sadiku et al., *Emerging Technologies in Manufacturing*,
https://doi.org/10.1007/978-3-031-23156-8_8

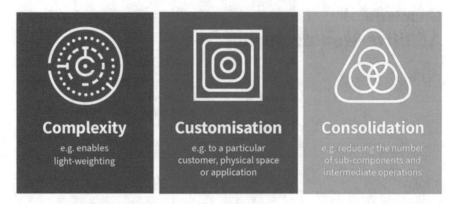

Fig. 8.1 Three components of additive manufacturing [2]

Fig. 8.2 A 3D printing machine at work [3]

8.2   What Is Additive Manufacturing?

Additive manufacturing (AM) technology first emerged in the 1980s. The rise of AM can be attributed to a technology called stereolithography that was developed in the 1980s. Additive manufacturing refers to a process by which 3D CAD data is used to construct an object in successive layers by depositing material. It is known as layer-upon-layer manufacturing. It may be regarded as a process whereby solid objects are constructed using additive techniques to lay down successive layers of a material until the object is completely made.

Additive manufacturing has some benefits over subtractive manufacturing. If a process is additive, it means that it is adding things. A subtractive process is one that is removing or taking away things. In subtractive or traditional manufacturing, you start with a large block of material and gradually cut away pieces of it until you get the desired object. In additive manufacturing, the manufacturing process adds instead of subtracts raw material. The material is added sequentially, layer upon layer. Fig. 8.3 compares subtractive and additive types of manufacturing [4].

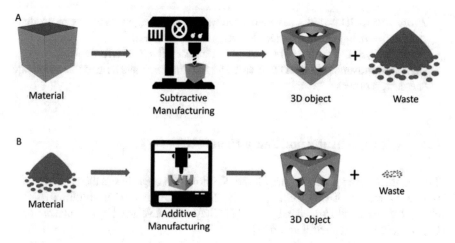

Fig. 8.3 Comparing subtractive and additive types of manufacturing [4]

Compared to subtractive manufacturing, additive manufacturing can compete with or beat traditional manufacturing processes in terms of cost and adaptability.

Additive manufacturing (AM) is also known as 3D printing, rapid prototyping, free form fabrication, or direct digital manufacturing (DDM). Although AM and 3D printing are not the same, both terms are used interchangeably. The term "3D printing" refers specifically to additive manufacturing processes that use a printer-like head for deposition of the material. AM is the industrial version of 3D printing, and it is growing at an incredible pace due to its potentially lower cost and flexibility.

AM technologies typically include a computer, 3D modeling software (computer-aided design or CAD), a machine, and material. The AM machine reads in data from the CAD file and develops successive layers of the material in a layer-upon-layer manner to fabricate the desired 3D object. The process involves using a computer and CAD software to relay messages to the printer so that it "prints" out the desired object.

AM is a digitally driven manufacturing process. The AM technology can take on any of the following five (5) distinct approaches [5]: (1) spray forming, using spraying equipment which is similar to the ink jet printer; (2) laminated object manufacturing, which consists of a computer, raw materials stored, and fed mechanism; (3) photosensitive polymer curing molding, which projects each layer image onto the liquid photosensitive polymer surface and solidifies every layer instantly; (4) materials extrusion molding, which builds the three-dimensional structure by means of accumulation, point to point, line by line, and layer by layer; and (5) laser powder sintering molding, which forms the required shape by bonding or fusing the powder material through thermal energy.

Additive manufacturing is really useful for the following general applications [6]:

- *Complex Shapes:* Many 3D printers can replicate fairly complex geometries and shapes.

- *Prototyping:* 3D printers have emerged as a prototyping method. It can produce prototypes using various materials with greater precision.
- *Availability:* 3D printing is becoming more accessible and affordable.
- *Software:* Advances in CAD software and 3D imaging have helped uncomplicate the design process.

8.3   Additive Manufacturing Processes

The manufacturing processes are classified into seven areas based the type of materials used, the deposition technique, and the way the material is solidified. These classifications are illustrated in Fig. 8.4 [7]. They were developed by the International Organization for Standardization [8]:

- *Powder Bed Fusion*: This process uses lasers, electron beams, or thermal print heads to melt ultrafine layers of material in a three-dimensional space. The powder bed fusion (PBF) technology is used in a variety of AM processes.
- *Binder Jetting:* This process uses a 3D printer style head that lays down and moves on the x, y, and z axes in order to alternate layers of material and a liquid binder. It can print a variety of materials including metals, sands, and ceramics.

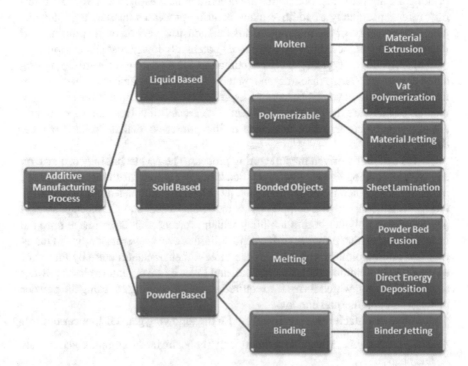

Fig. 8.4 Additive manufacturing processes [7]

Unlike other additive techniques, binder jetting does not heat during the build process and can print large parts.

- *Material Extrusion:* This is perhaps the most well-known additive manufacturing process. The nozzle moves horizontally, while the bed moves vertically, allowing the melted material to be built layer upon layer. Material extrusion is somewhat restricted in the types of shapes that can be created.
- *Directed Energy Deposition:* This process is similar to material extrusion, and it can be used with a wider variety of materials, including polymers, ceramics, and metals. It uses an electron beam gun or a laser that is mounted on a multi-axis arm.
- *Material Jetting:* In this process, a print head moves back and forth, much like the head on a 2D inkjet printer.
- *Sheet Lamination:* This process involves bonding sheets of material together layer by layer to form a single object. Laminated object manufacturing (LOM) and ultrasonic additive manufacturing (UAM) are two sheet lamination methods. LOM uses alternate layers of paper and adhesive, while UAM employs thin metal sheets conjoined through ultrasonic welding.
- *Vat Polymerization:* This process uses a vat of liquid photopolymer resin to construct an object layer by layer.

8.4 Applications

AM applications appear limitless. 3D printing or AM has been used in several domains such as medicine, construction, retail, defense, pharmacy, automotive industry, aerospace, electronics, food industry, engineering, education, architecture, and entertainment [5].

- *Aerospace:* The automotive and aerospace industries are the two main manufacturing beneficiaries of AM. AM technology is popular in aerospace/aviation industry because it can easily produce the lightweight and long-life components. In aerospace field, the rapid additive manufacturing of composites is currently the norm. AM is used in the aerospace industry to create hinges, brackets, interior components, and airframe designs in order to improve fuel efficiency. Aerospace companies are investing in AM technology to redesign parts to reduce weight, increase efficiency, reduce material waste, and combine parts—thereby reducing the number of items in complex assemblies. Hundreds of airline parts can be manufactured on an industrial 3D printer. It is particularly noteworthy and of some mild concern that aerospace would be an industry to embrace AM, given the tough industry performance standards which must be met as aerospace parts are often operating in some of the harshest conditions.
- *Automotive:* Additive manufacturing has enabled the automotive industry to create lighter and safer products. The automotive industry uses AM for rapid prototyping to experience reduced lead times and a reduction in costs. AM can be

found in various finished vehicle products such as in battery covers, air conditioner ducts, spare parts, and front bumpers.

- *Medicine:* The main applications of AM technology in medical/healthcare industry include orthopedics, plastic surgery, surgical implants, tissue engineering, and regenerative medicine. AM is used for fabrication of orthopedic implants, biomaterials, tissues, and organs. AM can construct patient-matched implants. It reduces the cost of production, development, and cycle time and enables rapid product fabrication. Advances in printable biomaterial and AM technologies allow the fabrication of vascularized tissue constructs. AM also makes operations faster, cost-efficient, and more accurate than manual processes. A few items that AM is used to create are dental prosthetics such as crowns, bone defects repair through customized implants, and custom fit masks [9].

- *Food Industry:* AM has two key strengths, geometric complexity and economy at low volume of production, and these translate into food applications. Its ability to produce items in small batches allows for customization. The key motivators for 3D printing of food products are customization, on-demand production, and geometric complexity. For additive manufacturing to be successful, it must integrate easily with traditional food production [10]. AM is being used in the food industry for squeezing out food, layer by layer, into three-dimensional objects. This applies to a variety of foods such as candy, crackers, pasta, and pizza.

- *Electronics:* Additive manufacturing has the potential to accelerate the pace of electronics manufacturing since it enables faster prototyping. It has been applied in fabricating active electronic components such as transistors, light-emitting diodes (LEDs), batteries, and operational amplifiers. (Transistors are important electronic components used in virtually all electronic devices.) These components usually require highly elaborate fabrication processes due to their complex functionalities [11]. Today AM is used for mass production of consumer electronics devices.

- *Military:* The military is getting more interested in additive manufacturing. It could offer them an efficient way to replace components when necessary. The US Department of Defense is considering additive manufacturing technologies to rapidly prototype and build equipment components. The US Naval Research Laboratory uses 3D printing to develop items for radar technology. With 3D printing, there exists a proven and reliable system on naval ships to make devices or replacements on an ad hoc basis.

- *Automation:* Automation can help maximize operational efficiencies, drive lean practices, eliminate redundancies, and simplify manufacturing processes. To automate AM technologies, companies are now beginning to implement robotic solutions to automate production. According to this concept, a robot will handle most of the process steps, such as feeding the printer with build boxes and then removing them for post-processing. The goal is to eliminate all manual work to facilitate continuous, high-volume production.

- *Energy:* The success of AM in the energy sector is similar to that it has enjoyed in the aerospace industry. As it enables the rapid development of relatively high strength, low weight custom-designed components are able to withstand extreme conditions.

Other areas of application include fashion designing, passive microwave components such as waveguides, couplers, power dividers, filters, and antennas), architectural design, civil engineering, and weapons development.

8.5   Benefits

Additive manufacturing technologies differ from traditional or subtractive manufacturing technologies. They have some advantages and disadvantages. They are changing both how and what can be manufactured. The strengths of additive manufacturing lie where conventional manufacturing reaches its limitations.

The demand in the global 3D printing market is gaining traction due to the following reasons [12]:

- The costs of 3D printers are falling.
- The geometric limits imposed by subtractive manufacturing no longer exist.
- It is possible to create multiple versions of the same product.
- A drastic reduction of processing waste is obtained.
- Strikingly higher resolution.
- Ease in the development of customized products.
- Growing possibilities of using multiple materials for printing.
- Government investments in 3D printing projects.
- Lower energy consumption.
- Less waste.
- Less dedicated tooling.
- Reduced development costs and time to market.
- Innovative designs and geometries.
- Part consolidation (fewer parts with more complex design).
- Customization of parts (e.g., for medical implants, spare parts).

Other benefits of additive manufacturing include the following [13]:

- *Flexibility:* Perhaps the most important benefit is their high design flexibility. The greater range of shapes which can be produced. AM enables a design-driven manufacturing process. It provides a high degree of design freedom, resulting in the manufacture of small shape sizes at reasonable unit costs.
- *Prototyping:* Additive manufacturing has revolutionized production with rapid prototyping. It enables quick prototyping and reduces the lead time and cost of developing prototypes. AM technology has made it easier and affordable for small companies and individuals to develop a customized prototype. AM enables companies to do what otherwise would be very difficult or impossible.
- *Complex Shape Capability:* Additive manufacturing allows for highly complex structures. It allows one to produce the objects without machining, lathing, mill-

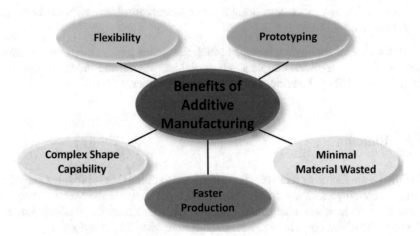

Fig. 8.5 Benefits of additive manufacturing

ing, grinding, boring, casting, or welding. AM has made possible the creation of objects that were previously very difficult or impossible to make with traditional manufacturing. The process is a faster way to make complex objects because the machines can run 24/7.

- *Faster Production:* Additive manufacturing offers production techniques that are faster than traditional ones. It is not a far-fetched idea to use AM in fabricating electronic devices in faster and in less expensive ways than current methods allow. Conventional methods such as injection molding can be less expensive for manufacturing polymer products, but additive manufacturing can be faster, more flexible, and less expensive.
- *Minimal Material Wasted*: A drastic reduction of processing waste is obtained. This makes AM more cost-effective, ultimately saving money.

Benefits are depicted in Fig. 8.5. Other benefits include green manufacturing, reduction in cycle time, speed to market, enabling personalized manufacturing, enabling design freedom, ability to eliminate joining, decentralized production, ability to build virtually any shape, and mass customization.

8.6 Challenges

There are issues that need to be addressed in order to make AM technology applicable for large-scale production. Understanding the difference between traditional and additive technologies is a challenge. AM systems have only become useful for mass production to a limited extent. It is still challenging to use AM on a larger scale. Industrial robots are often used along with AM technologies for producing large components. From industry viewpoint, the following eighteen barriers have been identified [14]: education, cost, design, software, materials, traceability, machine constraints, in-process monitoring, mechanical properties, repeatability, scalability, validation, standards, quality, inspection, tolerances, finishing, and sterilization. Some of these challenges are explained as follows [6, 15–19]:

- *Cost:* The greatest obstacle in implementing AM is the high price of the equipment. 3D printers are not cost-effective in most manufacturing situations. The needed materials can be expensive, so producing a large part may seem to be a waste of time.
- *Materials:* The raw materials used in additive manufacturing are still limited, mainly to plastics.
- *Low Speed:* Layers of material are gradually added to create the part in 3D printing. The larger the part, the longer the process takes.
- *Strength and Finish:* The layers of material create stratification. This process creates a rough surface finish, which must go through additional processing to obtain the desired feel.
- *Technical Limitations:* 3D printing is not fast enough to replace high-speed manufacturing processes, and it lacks flexibility in material use. The software needed to create a 3D file is still far too complex for an average person. 3D printing started with plastics and is still primarily done with plastics. AM with metals is more challenging due to their higher melting points and higher amount of energy required.
- *Security:* AM is a rapidly becoming a multibillion dollar industry. This makes AM an attractive attack target. There is substantial concern for the security of the storage, transfer, and execution of 3D models across digital networks. As additive manufacturing technology evolves so will the cybersecurity risk of tampering and misuse of designs. Security approaches for this emerging technology should to be addressed before the technology becomes more widely adopted.
- *Ownership and Copying:* This is important for intellectual property of designs. The question of ownership and copying of original ideas has sparked considerable debate and interest. AM offers new opportunities for counterfeiting products. For this reason, the government has introduced difficult-to-duplicate materials and designs into currency. The possibility of the rapid replication of highly sophisticated tools could vastly expand the number of persons able to commit violent acts or wage war.
- *Ethical Issues:* AM poses some ethical challenges for militaries and governments, who are responsible for keeping war-making tools out of the wrong hands. There is a current lack in the diversity of materials able to be processed using AM techniques. Other challenges include lack of industry standards, not suitable for mass production, and limited number of materials.
- *Lack of Standards:* The lack of 3D printing standards remains a major bottleneck slowing down its wider adoption. It is beneficial if the industry has standards which are universally understood and accepted. In manufacturing, standards are essential for raw materials, machines, equipment operators, engineers, suppliers, and the manufacturing process itself. Standards facilitate part design, and the

approval and certification of processes and standards promote adoption among engineers in the industry. Standards developing organizations, like ISO and ASTM, have issued a few specifications on metal powders like nickel, titanium, and stainless steel. *AM standardization is on a promising path.*

- *Lack of Skilled Labor:* AM requires engineers to develop a set of skills to support it; processes are required before and after. Currently, there are not enough AM engineers.

These shortcomings make AM several steps away from replacing current assembly lines of traditional manufacturing. As AM heads toward mass production, it needs to be environment-friendly and energy-efficient in order to be self-sustainable [20].

8.7   Global Adoption of Additive Manufacturing

Additive manufacturing is regarded as the next revolution in materials technology. It is receiving significant global attention. Much of the world continues to think of AM technology as a convenient way to make plastic objects from 3D printers. Several businesses across the world are now considering additive manufacturing for their supply chain. Some nations are using their own AM initiatives to help shape the manufacturing industry. Currently, the USA and Europe dominate the AM market, with some excellent work in Singapore, China, and Israel, among others. We consider how some regions adopt additive manufacturing.

- *United States:* In November 2016, the Department of Defense (DOD) released its first Additive Manufacturing Roadmap. The military personnel have been pushing its limits far beyond what most imagined possible. The Oak Ridge National Laboratory produced the military's first 3D-printed submarine hull [21]. GE uses key applications to showcase its AM offerings. The auto giant is already printing jet engine parts in commercial aircraft. Its additive manufacturing division specializes in developing powder bed fusion (PBF) machines for the additive manufacturing of metal parts.
- *United Kingdom:* In the United Kingdom, the AM-UK Steering Group has presented the AM-UK National Strategy, which includes brief summaries of the challenges facing industry. The strategy maps out how to overcome challenges in the following areas: cost/investment/financing, design, IP, protection and security, materials and processes, skills/education, standards and certification, and test and validation [14].
- *Germany:* The auto manufacturer BMW began its additive manufacturing of prototype parts in 1991. In 2019, it opened a center dedicated to additive manufacturing: in north of Munich. This center, shown in Fig. 8.6 [22], will facilitate the adoption of additive manufacturing in its activities, both for prototypes and series production. The German group has the *goal of increasing the industrialization of 3D printing techniques for automotive production while implementing new auto-*

Fig. 8.6 BMW additive manufacturing center [22]

mation concepts in the process chain. It hopes to become a leader in 3D printing technologies in the automotive sector. BMW is also involved in several projects supported by the German Ministry of Education and Research.

- *Southeast Asia:* Joint efforts by Asia Pacific Metalworking Equipment News, Siemens, Universal Robots, Markforged, and GlobalData are helping manufacturers understand 3D printing better and adopt it in Southeast Asia. They also examined the key enables and barriers toward successful adoption [23].

8.8 Conclusion

Additive manufacturing is regarded as a layer-upon-layer manufacturing. Although additive manufacturing is still in its infancy, it has the potential to revolutionize manufacturing. It will take manufacturing to the next level. It will lead to home manufacturing. It has been introduced in many companies worldwide. Some regard it as a game changer for the manufacturing industry. Do not allow your organization to be left in the dark ages by not adopting additive manufacturing.

Additive manufacturing is shaping the future of the manufacturing industry and becoming a mainstream manufacturing process. The field has a bright future. It provides a clear competitive advantage to customers [24]. Additive Manufacturing 2.0 is a wave of next-generation additive manufacturing technologies that will unlock throughput, repeatability, and competitive part costs. For more information on the AM field, one should consult the books in [7, 25–39] and the following international journals that exclusively devoted it:

- *Additive Manufacturing*
- *Progress in Additive Manufacturing*
- *Rapid Prototyping Journal*

These journals will help the reader to stay on top of the exploding field of additive manufacturing.

References

1. M.N.O. Sadiku, S.M. Musa, O.S. Musa, 3D Printing in the chemical industry. Invention Journal of Research Technology in Engineering and Management, **2**(2), 24–26 (2018)
2. Additive manufacturing: Past, present and future, https://www.qualitymag.com/articles/96307-additive-manufacturing-past-present-and-future
3. 3D printing: Why is it called additive manufacturing? https://ecolink.com/info/why-is-it-called-additive-manufacturing/
4. What is additive manufacturing? Applications, technologies and benefits, https://bitfab.io/blog/additive-manufacturing/
5. L. Chen et al., The research status and development trend of additive manufacturing technology. Int J Adv Manufact Technol **89**(9-12), 3651–3660 (2017)
6. The whole truth about additive manufacturing & 3D printing, February 2015, https://www.globaltranz.com/additive-manufacturing-and-3d-printing/
7. K.R. Balasubramanian, V. Senthilkumar, *Additive Manufacturing Applications for Metals and Composites* (IGI Global, 2020)
8. What is additive manufacturing? https://www.ge.com/additive/additive-manufacturing
9. M. Javaid, A. Haleem, Additive manufacturing applications in medical cases: A literature based review. Alexandria J Med **54**, 411–422 (2018)
10. J.I. Lipton et al., Additive manufacturing for the food industry. Trends Food Sci Technol **43**(1), 114–123 (2015)
11. N. Saengchairat, T. Tran, C.K. Chua, A review: Additive manufacturing for active electronic components. Virtual Phys Prototyp **12**(1), 31–46 (2017)
12. Additive manufacturing aims to meet DOD's needs, April 2021, https://www.afcea.org/content/additive-manufacturing-aims-meet-dod%E2%80%99s-needs
13. K.S. Prakasha, T. Nancharaihb, V.V.S. Rao, Additive manufacturing techniques in manufacturing: An overview. Mater Today Proceed **5**, 3873–3882 (2018)
14. L.E.J. Thomas-Seale et al., The barriers to the progression of additive manufacture: Perspectives from UK industry. Int J Product Econom **198**, 104–118 (2018)
15. S. Ivan, Y. Yin, Additive manufacturing impact for supply chain – Two cases, *Proceedings of the 2017 IEEE IEEM*, pp. 450-454.
16. S. M. Bridges et al., Cyber security for additive manufacturing, *Proceedings of the 10th Annual Cyber and Information Security Research Conference*, Oak Ridge, TN, April 2015.
17. M. Yampolskiy et al., Security of additive manufacturing: Attack taxonomy and survey, *Addit Manuf*, 2018 (to be published).
18. T. Kurfess, W.J. Cass, Rethinking additive manufacturing and intellectual property protection. *Res Technol Manag* **57**(5), 35–42 (2014)
19. J.M. Mattox, Additive manufacturing and its implications for military ethics. J Mil Ethics **12**(3), 225–234 (2013)
20. S. Kumar, A. Czekanski, Roadmap to sustainable plastic additive manufacturing. *Mate Today Commun* **15**, 109–113 (2018)
21. C. Collins, Additive manufacturing, October 2019, https://www.defensemedianetwork.com/stories/additive-manufacturing-department-of-defense-3d-printing-military-logistics/

22. V. Carlota, BMW is opening an additive manufacturing centre to pool its expertise, June 2020, https://www.3dnatives.com/en/bmw-additive-manufacturing-centre-290620205/#!
23. Recap: Additive manufacturing deployments in Southeast Asia, November 2020., https://www.equipment-news.com/recap-additive-manufacturing-deployments-in-southeast-asia/
24. The future of additive manufacturing in engineering, https://www.nano-di.com/blog/2019-the-future-of-additive-manufacturing-in-engineering
25. B. Badiru, V. Valencia, D. Liu (eds.), *Additive Manufacturing Handbook* (CRC Press, Boca Raton, FL, 2017)
26. A. Gebhardt, *Manufacturing:3D Printing for Prototyping and Manufacturing* (Verlag, Munich, 2015)
27. D.R. Gibson, B. Stucker, *Additive Manufacturing Technologies: 3D Printing, Rapid Prototyping, and Direct Digital Manufacturing*, 2nd edn. (Springer Science+Business Media, New York, 2015)
28. A.R. Pou, P. Davim (eds.), *Additive Manufacturing* (Elsevier, 2021)
29. J.K. Gebhardt, L. Thurn, *3D Printing: Understanding Additive Manufacturing*, 2nd edn. (Hanser Publications, 2018)
30. K. Chua, K.F. Leong, *3D Printing and Additive Manufacturing: Principles And Applications*, 5th edn. (WSPC, 2016)
31. Bandyopadhyay, S. Bose (eds.), *Additive Manufacturing*, 2nd edn. (CRC Press, Boca Raton, FL, 2019)
32. T.S. Srivatsan, T.S. Sudarshan, *Additive Manufacturing: Innovations, Advances, And Applications* (CRC Press, Boca Raton, FL, 2015)
33. Gu, *Laser Additive Manufacturing of High-Performance Materials* (Springer, 2015)
34. M. Leary, *Design for Additive Manufacturing* (Elsevier, 2019)
35. O. Milewski, *Additive Manufacturing of Metals: From Fundamental Technology to Rocket Nozzles, Medical Implants, and Custom Jewelry* (Springer, 2017)
36. AlMangour (ed.), *Additive Manufacturing of Emerging Materials* (Springer, 2019)
37. M. Devine (ed.), *Polymer-Based Additive Manufacturing: Biomedical Applications* (Springer, 2019)
38. R. Singh, J.P. Davim (eds.), *Additive Manufacturing: Applications and Innovations* (CRC Press, Boca Raton, FL, 2018)
39. G.K. Awari et al., *Additive Manufacturing and 3D Printing Technology: Principles and Applications* (CRC Press, Boca Raton, FL, 2021)

Chapter 9
Green Manufacturing

> *Meeting the needs of the present without compromising the ability of future generations to meet their own needs.*
> –United Nations

9.1 Introduction

Manufacturing has been a major source of resource consumption and environmental pollution. It is well-known as the largest sector of the American economy. It is closely connected with all other sectors like mining, trading, transportation, supply chain, and financial services. It contributes to the economy by providing many job opportunities, creating wealth, eradicating poverty, and providing better life standards, healthcare, and education [1]. However, the rapid technological advancements have led to a growing concern for environmental degradation caused by the manufacturing sector. The manufacturing sector accounts for a significant portion of the world's consumption of resources and generation of waste. It has negatively impacted the environment through the over-exploitation of natural resources and pollution. The manufacturing industry's energy demand is one-third of the total energy consumption in the United States [2]. To minimize the environmental damage due to manufacturing requires a new manufacturing process.

Green manufacturing (GM), also known as environmentally conscious manufacturing, is a new trend for the future development of the manufacturing industry. It is the embodiment of the strategy for sustainable development of the manufacturing sector. Green manufacturing refers to modern manufacturing that makes products without pollution. It alleviates the current contradiction between industrial development and environmental degradation and pollution. It addresses a wide range of environmental and sustainability issues including resource selection, transportation, manufacturing process, and pollution. This new way of thinking about manufacturing is having a big impact on manufacturers worldwide as they realize the many financial benefits of adhering to sustainable principles [3, 4].

This chapter provides a brief introduction to green manufacturing, an area of great importance for current and future manufacturing operations. It begins with traditional manufacturing concepts and then describes what GM is all about. It

© The Author(s), under exclusive license to Springer Nature Switzerland AG 2023
M. N. O. Sadiku et al., *Emerging Technologies in Manufacturing*,
https://doi.org/10.1007/978-3-031-23156-8_9

addresses the drivers and characteristics of GM. It covers green manufacturing practices. It highlights some applications of GM and discusses how to improve awareness about GM. It covers the benefits and challenges of GM. The last section
concludes the chapter.

9.2 Traditional Manufacturing

Manufacturing is the way of transforming resources into products or goods which
are required to cater to the needs of the society. It started from a small-scale production line of crafts in the 1800s. It has evolved into large-scale mass production. The
2000s is the era when computerized and personalized manufacturing systems came
into existence for mass production lines [5]. As illustrated in Fig. 9.1 [6], manufacturing is typically a series of processes comprising of selection of raw materials,
production of objects, assembling of parts, inspection, and dispatching. It may
include foundry, forging, jointing, heat treatment, painting, etc. It involves using
resources to meet the delivery date, cost, quality, and optimal economic goals in
limited resources condition. It often involves mass production and heavy energy
consumption such as coal or electricity.

The traditional manufacturing processing techniques consume a lot of energy,
mainly for production and utility. They also produce a lot of pollution and add to the
deterioration of the global environment. Because of this, manufacturers are gradually transforming their manufacturing systems from traditional mass production to
flexible lean systems. With rapid changes in technology, manufacturing itself is constantly transforming and evolving, as illustrated in Fig. 9.2 [7]. It now takes a proactive role in the development of cleaner manufacturing processes. In order to
minimize the environmental damage due to manufacturing, there is a need for new
manufacturing processes.

Fig. 9.1 Stages in the
manufacturing process [6]

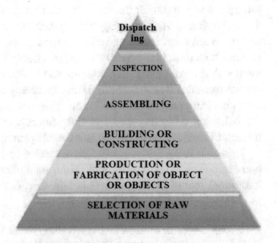

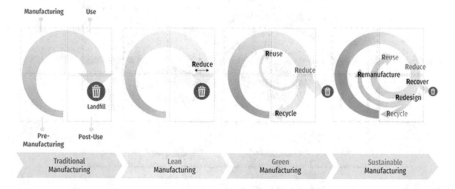

Fig. 9.2 Evolution of manufacturing [7]

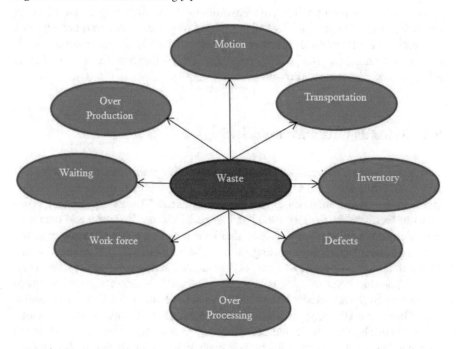

Fig. 9.3 Different types of production wastes [8]

In modern times, two types of manufacturing systems have emerged emphasizing waste minimization. They are "lean" manufacturing systems and "green" manufacturing systems that both reduce waste. Fig. 9.3 shows different types of production wastes [8]. Lean manufacturing seeks to eliminate all types of wastes generated within a production system. In lean manufacturing, there are eight categories of waste that should be monitored [8]: (1) overproduction, (2) waiting, (3) inventory, (4) transportation, (5) overprocessing, (6) motion, (7) defects, and (8) workforce. Lean and green best practices are considered complementary as shown in Fig. 9.4 [9]. Lean manufacturing is transcending to a more green state.

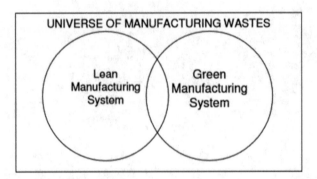

Fig. 9.4 Lean and green best practices are complementary [9]

Green manufacturing is different from traditional manufacturing in that it focuses on environmental impact, environment policies of governments, and national and international environmental regulations. Green manufacturing is an advanced, modern manufacturing approach that comprehensively considers the environmental influence and the resources consumption.

9.3   What Is Green Manufacturing?

Green stands for ecological sustainability. The term "green manufacturing" (GM) was coined to reflect a new manufacturing paradigm which implements various green strategies and methods to become eco-efficient. The concept of GM originated in Germany in the late 1980s and early 1990s, but its scope and nature of activities have been changing. GM is also known as lean manufacturing, environmental consciousness manufacturing, or no waste manufacturing. The term GM can be considered at in two ways. First, GM involves manufacturing of "green" (sustainable) products. Second, GM reduces emission and minimizes waste. Thus, green stands for ecological sustainability and GM is at the center of sustainability revolution. Green manufacturing is a sustainable approach that involves using energy, water, materials, and other resources efficiently to achieve balance with nature. As shown in the green manufacturing process cycle in Fig. 9.5 [10], GM is about reducing or eliminating any negative impact on the environment by a company's manufacturing facilities. It involves manufacturing practices that do not harm the environment during the manufacturing process [10].

Green manufacturing is a philosophy rather than a process. It is an important aspect of a circular economy. The objective of green manufacturing is to reduce environmental waste and pollution. It can be applied in all manufacturing sectors that minimize waste and pollution and conserve resources. Green manufacturing is an effective way to protect resources and the environment and conserve the resources for future generation. It is a key step toward achieving the big goal of sustainability.

Fig. 9.5   Green manufacturing process cycle [10]

9.4   Motivations for Green Manufacturing

In order to stay competitive, manufacturing companies are expected to implement green manufacturing and increase product complexity. This is becoming essential in the eyes of customers, investors, and authorities. Failure to comply may lead to fines, penalties, and customers choosing to go to the competitors.

For conceptual and common drivers of the philosophy of green manufacturing, different terms, such as "critical indicators," "critical success factors," "enablers," "focus areas," "motives," and "motivators," are used synonymously. The twelve common drivers of GM include financial benefits, company image, environmental conservation, compliance with regulations, stakeholders, green innovation, supply chain requirements, customers, employee demands, internal motivations, market trends, and competitors. Commitment from the top-level management is the most critical factor in the implementation of green manufacturing [11].

The manufacturing industry is motivated to adopt green manufacturing practices to reduce environmental impact and improve economic performance. Such practices include pollution prevention, product stewardship, and emission control. There are a number of factors that motivate the manufacturing industry to implement green manufacturing. These factors can be grouped into three categories of regulatory pressure, economic incentives, and competitive advantages. These factors include [12]:

- Pressure from the government—regulations and tax benefits
- Access to government incentives
- Increase sales

- Save money on energy costs
- Boost employers' morale
- Interest in efficiency
- Scarcity of resources
- Pressure from society/consumers and competitors
- Desire to maintain market leadership
- Ensure control of supply chain effects

US governmental agencies have developed a series of policies, regulations, and laws, which have achieved significant progress in advancing the environmental performance of manufacturers. These governmental efforts compel the manufacturing industry to consider green manufacturing as the economic benefits which could result from the implementation of sustainability programs. For example, the US Food and Drug Administration (FDA) has the responsibility for evaluating the safety and efficacy of new drugs.

International organizations such as the United Nations, World Bank, and the Carbon War Room are making notable progress to promote green manufacturing. Stakeholders, customers, and the government are increasingly asking companies to be more environmentally responsible with respect to their products and processes. Their reasons for this requirement include regulatory requirements, product stewardship, public image, and potential competitive advantages.

9.5   Characteristics of GM

The four "Rs" (reduce, reuse, recycle, and remanufacturing) are one main strategy of green manufacturing. The 4R principles of green manufacturing are illustrated in Fig. 9.6 and explained as follows [13].

- *Reduce:* This requires decreasing the consumption of resources (e.g., energies, water, materials) and emission of wastes including atmospheric pollutions, e.g., carbon emissions, carbon monoxide, carbon dioxide, sulfur dioxide, and photochemical smog.
- *Reuse:* This requires reusing the products with the aim of prolonging the life of the products and reducing waste.
- *Recycle*: There are two ways of recycling. One of which produces the same kind of new products, while the other is when recycling is transferred into raw materials of other products.
- *Remanufacturing:* This is an approach to recover the old products back to the ones close to the new products.

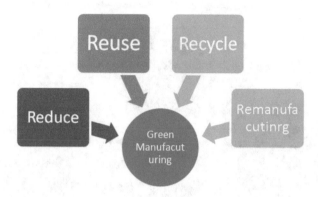

Fig. 9.6 The 4R principles of green manufacturing [13]

9.6 Green Manufacturing Practices

The term "green manufacturing" originated in Germany in the late 1980s. Green manufacturing is involved in the whole product lifecycle. It is a step toward sustainable manufacturing. As illustrated in Fig. 9.7 [14], green manufacturing has the intent of addressing all five areas: green resources, green design, green production, green manufacturing, and green disposal.

- *Green Resources*: Green manufacturing uses energy, water, materials, and other resources more efficiently thereby reducing the overall impact on the environment. Workers use fewer natural resources. Using fewer natural resources to make the same product saves money. Workers may generate electricity from renewable sources which include wind, biomass, geothermal, solar, ocean, and hydropower. Green energy may be regarded as an environmental strategy, a national security strategy, or an economic strategy.
- *Green Design:* This is also known as environment-conscious design or product lifecycle design. Green design focuses on developing eco-friendly products that minimize waste. It considers the environmental factors and measures pollution prevention in the product design phase. Product designer needs to consider the manufacturability of the product, the energy consumption, the maintainability, and the reusability of the product [15]. If environmental protection is considered at the design stage of a product, pollution can be easily reduced to the minimum. Manufacturing companies need to shift toward using cleaner energy.
- *Green Production:* Green technologies focus on reducing the impact of manufacturing processes at every stage of production. Companies should stay ahead of the curve on sustainability and evaluate how green are the products, technologies, resources, and energy. The selection of raw materials directly dictates the realization of green production. A growing range of green products, ranging from organic food products to electric cars, are being offered by companies to customers [16]. By developing green products that are demanded by consumers, companies can derive additional sales that can offset their cost of development. Green

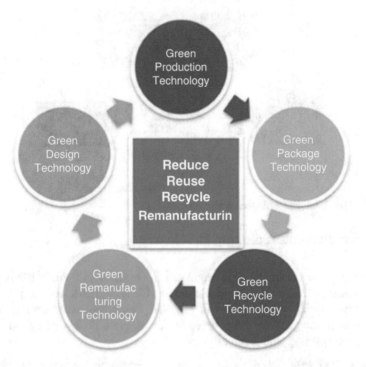

Fig. 9.7 Principles of green manufacturing [14]

products save energy, reduce energy consumption, and increase the company's competitive advantage.

- *Green Manufacturing*: Green manufacturing considers the impact of product development, manufacturing, and activity on the environment. This assures sustainability in resource extraction, material processing, product use, and disposal. The green manufacturing process obeys the following five principles [17]: (1) least resources consumption principle, (2) least energy consumption principle, (3) least environment pollution principle, (4) better labor protection principle, and (5) economic efficiency principle.

- *Green Disposal:* Green disposal aims to reduce e-waste by repairing, redeploying, disposing, refurbishing, retaining, and reusing of outdated IT hardware. Reducing resource use, waste, and pollution, along with recycling and reusing waste, yields benefits. Waste is whatever does not add value to the end product. To achieve green disposal, the release of toxic substances in product life is not allowed. Overproduction, which is a form of waste, is also disallowed, which occurs when production output exceeds actual customer orders.

9.7 Green Manufacturing Strategies

Green manufacturing is decidedly a philosophy for manufacturing that minimizes waste and pollution through product and process design. The main objective of green manufacturing is sustainability. Sustainable trend is becoming crucial in all aspects of manufacturing. Today, manufacturers must keep the environment in mind at each step of their product lifecycle. To go green, one should consider the following manufacturing strategies [18]:

1. Design for disassembly, remanufacture, or reuse.
2. Rethink product and process technology.
3. Streamline the supply chain.
4. Reduce energy and water consumption.
5. Choose recyclable or biodegradable materials and packaging.
6. Integrate environmental costs into the organization's production budget.
7. Use ISO 14001 standard as a jumping-off point.
8. Find a reverse logistics supply vendor.
9. Invest in business intelligence/analytics.
10. Redesign all scales of manufacturing flow.
11. Shift to a service-oriented business.

Some organizations have started developing competitive GM strategies. This enhances the image of companies in the eyes of the customer and their competitiveness. Green manufacturing strategic patterns adopted by ISO 14001 certificate holders in Jordan were agile, lean, and caretaker strategic patterns. The green strategic patterns are the milestones of the green practices' success. The effectiveness of adopted green strategies is yet to be determined [19].

9.8 Applications of Green Manufacturing

Green manufacturing refers to the new manufacturing paradigm that employs various green strategies and techniques to become more eco-efficient. It is a modern manufacturing approach that gives due consideration to environmental impact and resource consumption. It seems to be an effective way to achieve sustainable manufacturing development. This new green manufacturing paradigm is an outcome of market and technological drivers. It may also be regarded as an important component of green business. Additive manufacturing or 3D printer technologies needs to be considered as a practical green process because of saving in materials and reduction in processing steps.

Interest in green manufacturing is increasing more and gaining traction within industrial communities [20]. The following are typical applications of GM across a broad range of industries [21].

- *Automotive Industry:* The automotive industry has become indispensable, and it plays a vital role in the national economy. The industry is based on the great consumption of resources and energy to meet the mobility needs of the masses, while the continuous and incremental disposal of end-of-life vehicles leads to severe pollution [13].
- *Semiconductor Industry:* The production of semiconductor/electronic products (such as instruments, radio, TV, radars, computers, mobile phones) consumes a great deal of resources as well as generates harmful wastes during their production.
- *Iron and Steel Industry*: This is an essential industry in the national economy, but it consumes raw materials to produce iron and steel. Steel is the world's leading industrial product. Traditionally, the steel industry produces iron oxide and carbon, leading to large carbon emissions. One way to improve environmental friendliness of iron and steel is to deploy source control strategies with the intent of reducing iron and steel consumption by-products of industrial and atmospheric pollution. Green manufacturing can be adopted to manage iron and steel production [22].
- *Pharmaceuticals Industry:* The pharmaceuticals industry fabricates chemical products and receives strict regulatory oversight to protect the public from unsafe pharmaceutical products. For example, paint and pigment manufacturing industries generate large quantities of hazardous and nonhazardous wastes resulting in ill-health effects and chronic health conditions including tumorous and cancerous effects. Reducing waste should be a high priority for this manufacturing sector.
- *Transportation:* Transportation consumes much of the earth's resources. It is key to greening global industries. A basic change in vehicle design, including powertrain and propulsion systems design, noise, and emissions control, is necessary. It will reduce emissions, fuel consumption, and cost. Electric vehicles, fuel cells, and biodiesel are some examples of this category. Governments around the world are tightening the regulations on emissions caused by transportation.
- *Textile Industry*: This industry gathers the agricultural, chemical fiber, textile, apparel, retail, service, and waste management sectors. Antibacterial textile structure is used as a potential replacement for synthetic fabrics or cotton fabrics. Carpet industry is an early environmental mover within the textile industry. It is advanced with respect to sustainable manufacturing practices [23, 24].
- *Computer Industry:* Apple is optimizing the energy efficiency of its hardware tools. Motorola improves recyclability, minimizes packing, and reduces hazardous materials. All Samsung vendor facilities desire to be 1SO 14,000 compliant.
- *Cement Industry:* Cement manufacturing uses a lot of energy and produces air emissions like nitrogen oxides, sulfur oxides, carbon dioxide, hydrochloric acid vapor, and chlorine. The pollution from the cement industry will keep increasing because of the increased demand for new houses. There is a need for controlling the environmental concerns raised by the cement industry through green manufacturing practices adoption [25].

- *Other Companies:* Other companies that have taken similar measures include GM, Ford, IBM, IKEA, Dell, Johnson & Johnson, and Nike. WalMart, the world's largest retailer, is betting on hybrid technology to power its truck fleet.

Other applications that could benefit from the green revolution include renewable energy systems, green chemistry avoidance of toxins, furniture industry, textile industry, carpet industry, foundry industry, air products industry, and the oil and gas industry.

9.9   Awareness of GM

The awareness of GM is crucial to its adoption. In order to adopt and implement GM, it is important for society to be aware of its significance. There is no doubt that GM can be easily accepted and supported by the populace, the government, and nonprofit organizations. Awareness about green manufacturing must be improved through education. Everyone associated with the business must be involved in adopting green manufacturing, including suppliers, customers, and employees. Training and education programs are thus essential for employees. Students and workers in manufacturing must learn about green manufacturing, sustainability concepts, and practices such as [26]:

1. *Energy from Renewable Sources:* Workers may generate electricity from renewable sources such as wind, biomass, geothermal, solar, ocean, hydropower, etc.
2. *Energy Efficiency.* Workers will utilize specific technologies and practices to improve energy efficiency. More energy-efficient technologies will help in using renewable energy and protect the world from climate change.
3. *Pollution Reduction:* Workers will use green technologies and practices to reduce the generation of pollutants. Manufacturers must take all reasonable steps to eliminate pollution.
4. *Natural Resources Conservation:* Workers will use specific technologies and practices to conserve natural resources such as land, water, etc. With resources becoming scarcer due to depletions and exploitations, it is imperative to reduce waste by minimizing natural resource use and recycling.

Organizations can assess the commitment of the top management to GM in terms of the availability of an explicit environmental policy, availability of effective strategies to achieve the policy, and allocation of financial resources [27].

9.10   Benefits

GM is regarded as the winning strategy to be adopted by manufacturers worldwide.
Successful adoption of green manufacturing will bring many benefits (economic benefits, environmental benefits, and social benefits) both tangible and intangible

for each nation and the international business community. Manufacturing companies will adopt green manufacturing if they realize that it will result in some financial benefits. Green manufacturing can mitigate air, water, and land pollution. It reduces waste at the source and minimizes health risks to humans. Other benefits of green manufacturing include [28–31]:

- *Reduce Costs*: The costs associated with manufacturing can be reduced through sustainability initiatives. Green manufacturing can be practiced by reducing the cost of raw materials, which is achievable by using less energy and reusing recycled wastes instead of buying new materials for production [2]. Perhaps the largest payoff of the green manufacturing approach is energy savings since energy costs are a major prime concern for manufacturers.
- *Reduce Consumption of Resources:* The manufacturing industry is responsible for significant resources and energy consumption. Green manufacturing helps to reduce resource use, waste, and pollution, along with recycling and reusing what was formerly viewed as waste.
- *Reduce Waste:* Waste consumes resources without adding any advantage to the product. Manufacturing does not have to be a wasteful or uneconomical process. Reducing industrial waste should be a top priority for manufacturing leaders, policymakers, and decision-makers in order to save our planet. An important aspect of environmental sustainability is reducing waste. The reduction in waste production delivers broad-based and far-reaching benefits more than the environment.
- *Reduce Pollution:* Manufacturing activities have been one of the major polluters of the environment. Green manufacturing primarily reduces or removes the release of pollutants in manufacturing operations, conserves energy and natural resources, and makes safer products for consumers.
- *Tax Incentives:* There are tax incentives at the federal and state levels for green manufacturers. The incentives can be leveraged to invest in new technologies.
- *Environmental Friendliness*: Every non-sustainable measure impacts the environment. A company can help the environment by simply optimizing existing operations and reducing agents of environmental degradation.
- *Adherence to Compliance Regulations*: Creating and implementing a sustainability strategy is the only long-term solution companies have, to avoid possible compliance violations. Increasing environmental demands from governmental agencies and customers emphasizes the need for companies to improve their environmental performance.
- *Energy Efficiency:* Green practices can be used to improve energy efficiency within an establishment. They minimize energy use in production.
- *Stability:* One of the best benefits of green manufacturing is that it improves the long-term stability of an enterprise. Companies are now starting to become more conscious of the environment, and so are customers. A great need exists, to align organizations with the new trends, in order to survive in the future.
- *Reputation Boost:* As consumers show interest in sustainability, launching businesses into the green manufacturing and clean energy arena can be a great repu-

tation boost that will increase overall business sales and profits. When companies go green, it shows in their customer response.

- *Societal Impact:* Going green can make a company more marketable. Future generations will benefit from improved air and water quality and more eco-friendly and renewable energy sources.
- *Increase Employee Morale:* GM yields benefits not only in terms of an improved bottom line, but in terms of employee motivation, morale, and public relations. When employees work together to implement green initiatives, it increases workforce morale and fosters a culture of teamwork.
- *Attracting New Employees:* One way to appeal to the next generation of workers is to demonstrate your commitment to the greening of the environment.
- *Attract New Customers:* Green practices can make your company more marketable, resulting in increased sales.
- *Global Demand:* Environmentally conscious manufacturing processes have become an obligation to the environment and to the society. Stakeholders including regulators, customers, shareholders, board members, and employees are increasingly demanding companies to be more environmentally responsible with respect to their products. Their reasons include regulatory requirements, product stewardship, enhanced public image, potential to expand customer base, and potential competitive advantages [23].

Some of these benefits are illustrated in Fig. 9.8. Other benefits of green manufacturing include lower raw material costs, production efficiency, and maintaining competitive advantage.

9.11 Challenges

In spite of the many benefits that green manufacturing offers to the environmental problems, it is facing some challenges including [32]:

- *Global Competition:* It can be difficult for companies to implement green manufacturing cost-effectively. It is challenging to compete with companies overseas that do not abide by the same standards and are therefore able to keep their operating costs low, thereby securing unfair competitive advantages.
- *Lack of Required Skills:* Green manufacturing jobs require a strong core skillset and an added layer of green knowledge. There is lack of required skills and shortage of financial resources.
- *Measuring Greenness:* The conventional manufacturing wisdom suggests that we cannot improve what cannot be measured. There is the challenge of measuring manufacturing greenness level. There is a need for developing metrics that track and monitor the performance of GM. A toolbox (Greenometer) to assess the greenness level of manufacturing companies has been proposed. Greenometer offers relative greenness assessment among different industries [33].
- Many companies are still skeptical about the business benefits of GM.

Fig. 9.8 Some benefits of green manufacturing

- Ascertaining the costs of embarking on green manufacturing.
- Lack of capable scientifically based decision support tools for effective implementation of GM.
- GM often requires the use of more expensive materials and technologies.
- The rate at which green manufacturing systems are being implemented is not keeping pace with the global expansion of the manufacturing industry.
- Assessing the performance of green manufacturing.
- Increase in production cost attributable to GM.
- The life cycle cost of product in green manufacturing is lower than that of the same product in traditional manufacturing.

Green manufacturing will have more benefits and challenges in the years to come.

9.12   Global Impact of Green Manufacturing

Green manufacturing essentially achieves two things: manufacturing of "green" products and "greening" of manufacturing, in which workers use fewer natural resources, reduce pollution, minimize waste, and recycle and reuse materials. It has attracted the attention of industries all over the world, and consumers are demanding it.

Manufacturing companies worldwide are now more conscious about greening the environment. The global manufacturing is no longer only concerned with the competition for sources, capital, and labor but also with the competitiveness of green technology. A lot of companies all over the world claim to be "going green." The United States and European Union are leaders in the implementation of environmentally conscious production. Examples of companies that are making strides in sustainable/green manufacturing include Coca-Cola, BMW, GM, Ford, Motorola, Toyota, Tesla, IBM, Dell, Siemens, Samsung, Apple, Nike, Johnson & Johnson, and

Tupperware. We now consider how different nations implement green manufacturing.

- *United States*: Over 70 percent of Americans regard manufacturing as the most important industry in the American economy. From retail to utilities, the manufacturing industry is closely connected with all other thriving industries across the nation. The United States remains the world's largest manufacturing economy, producing 21% of global manufactured products. A strong US manufacturing sector requires that the government policymakers enact stronger US government policies and incentives that promote and foster its ongoing development and growth. Industrial leaders have urged the government to enact policies and take actions to revitalize US manufacturing. By the same token, governments and nongovernment organizations have been urging manufacturers to clean up their act and take greater responsibility for resource use, waste, and pollution [3]. Manufacturers in the United States are turning toward sustainable alternatives to boost competitive advantage and increase revenue.
- *China:* China, as the biggest manufacturing nation, is taking great pains in practicing green manufacturing. The automotive industry has become one of the Chinese pillar industries. Although the industry creates wealth and job employment, it makes a significant impact on the environment. Manufacturing in China is facing severe challenges. China has been ramping up its green manufacturing initiatives. As China scales its manufacturing industries, it will create a large carbon footprint. Green manufacturing is one of the three development strategies for manufacturing industry in China. The adoption of green technology in the automotive industry has three aspects: green design, green processing, and green remanufacturing. The product design stage determines the type of materials and resources to be used. China is beginning to realize the concerns of global warming. China recently shut down 40% of its factories to minimize pollution. For example, in Beijing, 1000 manufacturing companies were shut down by 2020 as part of an initiative to reduce smog and waste. Beijing also plans to relocate all manufacturing and heavy industry companies. China may need outside technology to reach its ambitious sustainability goals. To establish a green university, green engineering has priority in the program developed by Tsinghua University, in the support of the US "China Bridge" [34–36].
- *Japan:* Sustainable product and process engineering, green, lean design, manufacturing, management rules, and principles are offered with a focus on "mono-zukuri." The Japanese term "monozukuri" means sustainable, environmentally friendly, green factories and products with simultaneously integrated product and process designs [37]. For example, Toyota's assembly plant generates green power from local landfill gas. The company's Prius has come to symbolize green motoring in parts of the world.
- *Europe Union:* Europe pioneered the manufacture of the smart car, which has experienced a surge in popularity among environmentally conscious consumers. Smart cars are made using eco-friendly production methods, resulting in less production-related emissions and waste. For example, BMW promotes its green

manufacturing programs. The company replaces high-solvent paints with water-based ones when possible. Environmental protection is high on the agenda of Mercedes-Benz. Other automakers such as Ford, General Motors, Honda, and Chrysler have followed suit.

- *India:* Manufacturing is an important sector for India since it creates both job and wealth. Green manufacturing in India is at the take-off stage. The Confederation of Indian Industry (CII) is a nongovernment, not-for-profit organization that creates and maintains an environment conducive to the growth of the industry in India. With several offices located within and outside India, CII serves as a reference point for the Indian industry and the international business community. The Indian foundry industry is known as the fifth in the world. The foundry sector consumes a lot of energy [38].
- *Ghana:* Like other nations, the manufacturing sector plays a major role in the economy of Ghana. It provides job opportunities to skilled job seekers and also serves as a source of huge foreign exchange inflows. However, the adverse effect of manufacturing activities on the environment cannot be underestimated. Industrial toxic substances being released in Ghana have been from the industries such as food processing, mineral exploitation, petroleum handling, and the textile industry. Ghana has been taking steps toward creating a green economy through national policies, blueprints, and initiatives. A significant effort made toward preventing the adverse effect of industrial activities on the environment includes the formation of the Environmental Protection Agency of Ghana (EPA), Environmental Impact Assessment (EIA), and the Switch Africa Green [39].

9.13   Conclusion

Green manufacturing has become the inevitable choice in the twenty-first-century manufacturing industry. It will increasingly be an important issue because it is a crucial component in the international effort toward "sustainable development." It is the only realistic means of realizing modern manufacturing sustainable development. For this reason, green manufacturing has received growing interest in the last few years. It is fast becoming a necessary business practice in the manufacturing industry. It is here to stay.

Green manufacturing will have a lot of benefits in the years to come. It is the important issue that manufacturing systems of the future must consider. There are various incentives, resources, and industry-specific organizations available to help manufacturing companies adopt green initiatives. As long as the benefits of being green outweigh the cost, the adoption of green practices will continue to be attractive and wise. More information about green manufacturing can be found in [12, 40–46] and the journals devoted to it:

- *International Journal of Precision Engineering and Manufacturing-Green Technology*
- *Journal of Engineering Manufacture*

References

1. D. Seth, M.A.A. Rehman, R.L. Shrivastava, Green manufacturing drivers and their relationships for small and medium (SME) and large industries. J Clea Prod **198**, 1381–1405 (2018)
2. J. Li et al., Editorial automation in green manufacturing. IEEE Trans. Autom. Sci. Eng **10**(1), 1–4 (2013)
3. M.N.O. Sadiku, U.C. Chukwu, A. Ajayi-Majebi, S.M. Musa, Green manufacturing: A primer. J Eng Sci Res Appl **7**(6), 134–137 (2020)
4. M.N.O. Sadiku, O.D. Olaleye, A. Ajayi-Majebi, S.M. Musa, Green manufacturing: An introduction. Int J Trend Res Dev **7**(3), 296–297 (2020)
5. Y. Nukman et al., A strategic development of green manufacturing index (GMI) topology concerning the environmental impacts. Procedia Eng **184**, 370–380 (2017)
6. D. Swathisri, D. S. S. Kumar, Green manufacturing technologies – A review, https://www.researchgate.net/publication/305731124_GREEN_MANUFACTURING_TECHNOLOGIES_-_A_REVIEW
7. https://www.engr.uky.edu/ism
8. G.D. Maruthi, R. Rashmi, Green manufacturing: It's tools and techniques that can be implemented in manufacturing sectors. Mater Today Proceed **2**, 3350–3355 (2015)
9. G.G. Bergmiller, *Lean manufacturers transcendence to green manufacturing: Correlating the diffusion of lean and green manufacturing systems.," Doctoral Dissertation,* (University of South Florida, 2006)
10. R.A.R. Ghazilla et al., Drivers and barriers analysis for green manufacturing practices in Malaysian SMEs: A preliminary findings. *Procedia CIRP* **26**, 658–663 (2015)
11. K. Govindan, A. Diabat, K.M. Shankar, Analyzing the drivers of green manufacturing with fuzzy approach. J Clean Prod **96**, 182–193 (2015)
12. D.A. Dornfeld (ed.), *Green manufacturing: Fundamentals and Applications* (Springer, New York, 2013), p. 8
13. Z. Zeya, *Green manufacturing framework development and implementation in industry, Master's Thesis,* (Tallinn University of Technology, 2015)
14. Z. Zeya, Green manufacturing framework development and implementation in industry, https://digi.lib.ttu.ee/i/file.php?DLID=3921&t=1
15. W. Qifen, Green manufacturing-oriented digital system and operation technologies for manufacturing enterprises. AMM **20-23**, 40–44 (2010)
16. A. Bhattacharya, R. Jain, A. Choudhary Green manufacturing: Energy, products and processes, 2011., http://www.cii.in/webcms/Upload/BCG-CII%20Green%20Mfg%20Report.pdf
17. D. Zhong, Study on a green manufacturing process design system. AMM **397-400**, 57–61 (2013)
18. Green manufacturing: 8 strategies for success, http://members.questline.com/Article.aspx?articleID=32753&accountID=1&nl=19072
19. Y.K.A. Migdadi, D.S.I. Elzzqaibeh, The evaluation of green manufacturing strategies adopted by ISO 14001 certificate holders in Jordan. Int. J. Product. Qual. Manag **23**(1), 90–109 (2018)
20. A.M. Deif, A system model for green manufacturing. Adv Prod Eng Manag **6**, 27–36 (2011)
21. S.H. Ahn, D.M. Chun, W.S. Chu, Perspective to green manufacturing and applications. IJPEM **14**(6), 873–874 (2013)
22. G. Li et al., Optimization of production in iron and steel enterprise for green manufacturing. Key Eng Mater **460-461**, 631–636 (2011)
23. C.A. Rusinko, Green manufacturing: An evaluation of environmentally sustainable manufacturing practices and their impact on competitive outcomes. IEEE Trans Eng Manag **54**(3), 445–454 (2007)
24. E. Alay, K. Duran, A. Korlu, A sample work on green manufacturing in textile industry. Sustain Chem Pharm **3**, 39–46 (2016)
25. S.M. Eshikumo, S.O. Odock, Green manufacturing and operational performance of a firm: Case of cement manufacturing in Kenya. Int J Bus Soc Sci **8**(4), 106–120 (2017)

26. What is green manufacturing and why is it important? https://www.goodwin.edu/enews/what-is-green-manufacturing/
27. P. J. Singh, K.S. Sangwan, Management commitment and employee empowerment in environmentally conscious manufacturing implementation, *Proceedings of the World Congress on Engineering,* London, U.K., July 2011
28. Five benefits of sustainability and green manufacturing, https://www.qualitydigest.com/inside/customer-care-column/101816-five-benefits-sustainability-and-green-manufacturing.html
29. Top 5 benefits of green manufacturing, http://www.impactdakota.com/blog/top-5-benefits-of-green-manufacturing
30. C. Y. Jian, The role of green manufacturing in reducing carbon dioxide emissions, *Proceedings of the Fifth Conference on Measuring Technology and Mechatronics Automation,* 2013, 1223–1226.
31. A. Robinson, Green is the new black: Why green manufacturing & sustainability matter, May 2017. https://cerasis.com/2017/05/17/green-manufacturing/
32. Annie Qureshi, Challenges and opportunities of green manufacturing, February 2018, https://blueandgreentomorrow.com/news/challenges-opportunities-green-manufacturing/
33. A.H. Salem, A.M. Deif, Developing a Greenometer for green manufacturing assessment. J Clean Prod **154**, 413–423 (2017)
34. Renewable energy & clean technology: Keys to a revitalization of US manufacturing & job creation, April 2012, Unknown Source.
35. P. Li, H. Zhan, Application situation and development strategies of green manufacturing in China's automotive industry. Adv Mat Res **1049-1030**, 945–948 (2014)
36. Y. Mao, J. Wang, Is green manufacturing expensive? Empirical evidence from China, *International Journal of Production Research,* 2018.
37. P. G. Ranky, Sustainable green product design and manufacturing / assembly systems engineering principles and rules with examples, *2010 IEEE International Symposium on Sustainable Systems and Technology,* May 2010.
38. S. Gupta et al., Implementation of sustainable manufacturing practices in Indian manufacturing companies. Benchmarking Int J **25**(7), 2441–2459 (2018)
39. W. Gyasi-Mensah, H. Xuhua, Towards developing a green manufacturing environment: What is Ghana doing? Env Manag Sustain Dev **7**(2) (2018)
40. N.K. Jha, *Green Design and Manufacturing for Sustainability* (CRC Press, Boca Raton, FL, 2015)
41. M.J. Franchetti, B. Elahi, S. Ghose, *Value Creation Through Sustainable Manufacturing* (Peen Tool Co, Maplewood, NJ, 2016)
42. J.K. Wang, *Green Electronics Manufacturing: Creating Environmental Sensible Products* (CRC Press, Boca Raton, FL, 2017)
43. M. Singh, T. Ohji, R. Asthana, *Green and Sustainable Manufacturing of Advanced Material* (Elsevier, 2015)
44. G. Seliger (ed.), *Sustainable Manufacturing: Shaping Global Value Creation* (Springer, 2012)
45. M. Uthayakumar et al. (eds.), *Handbook of Research on Green Engineering Techniques for Modern Manufacturing* (IGI Global, 2018)
46. J. Rynn, *Manufacturing Green Prosperity: The Power to Rebuild the American Middle Class* (Praeger, Santa Barbara, CA, 2010)

Chapter 10
Sustainable Manufacturing

> *Sustainability is not only central to business strategy, but will increasingly become a critical driver of business growth. How well and how quickly businesses respond to this agenda will determine which companies succeed and which will fail.*
> —Patrick Cescau

10.1 Introduction

Manufacturing is the main pillar of the modern society. It is a well-acknowledged fact that traditional manufacturing processes are generally designed for high performance and low cost with little attention paid to environmental issues. Although manufacturing systems create material wealth, they consume a great amount of resources and generate a lot of waste, which is responsible for the degradation of the environment. Thus, such traditional manufacturing processes have retained the negative image of being inefficient, polluting, and harmful. Manufacturers are under pressure by regulations and consumers to reduce the environmental impact of their activities.

Today, humans are consuming natural resources through manufacturing activities at an alarming rate, which is not sustainable. For example, between 1950 and 2005, worldwide metals production grew sixfold, oil consumption eightfold, and natural gas consumption 14-fold. The current assumption of unlimited resources and unlimited world's capacity for regeneration is no longer acceptable. Thus, minimizing the resource consumption and reducing the environmental impact of manufacturing systems has become very important [1].

Sustainable manufacturing (SM) is manufacturing products through economically sound processes that minimize negative environmental impacts while conserving energy and natural resources. The goal of sustainable manufacturing is to minimize waste, maximize resource efficiency, and reduce the environmental impact of manufacturing. It is imperative that manufacturing processes should consider sustainability at every level, so that there will be comprehensive adherence to sustainability principles. Properly implemented, sustainable manufacturing can lead to several advantages.

This chapter describes sustainable manufacturing and how environmental sustainability helps in achieving it. It begins by explaining the concept of sustainability.

© The Author(s), under exclusive license to Springer Nature Switzerland AG 2023
M. N. O. Sadiku et al., *Emerging Technologies in Manufacturing*,
https://doi.org/10.1007/978-3-031-23156-8_10

It discusses what sustainable manufacturing is all about. It covers some sustainable approaches. It provides some examples of sustainable manufacturing. It highlights the benefits and challenges of SM. It covers the global adoption of SM. The last section concludes with comments.

10.2   The Concept of Sustainability

Sustainability is the future of manufacturing. It has been a keyword in the twenty-first century because it is one of the global grand challenges. For example, we hear about sustainable engineering, sustainable development, sustainable energy, sustainable software, sustainable design, sustainable living, economic sustainability, social sustainability, ecological sustainability, etc. In this same way, there has been considerable discussion about green chemistry, green engineering, green business, green manufacturing, green food, green economy, green energy, etc. The two terms (sustainability and green) are often used interchangeably [2]. Sustainable development has been a major driving initiative in engineering businesses throughout the world. Green engineering involves creating healthy living environments that use natural resources wisely and conservatively [3].

Sustainability requires monitoring your operation and continually looking for ways to improve. It has been applied to many areas including manufacturing, engineering, design, and business. It will be a crucial issue for the present and future generation of manufacturing solutions. When we talk about sustainability, we often mention the three Ps: people, planet, and profit, as shown in Fig. 10.1 [4]. The pillars of sustainability are a powerful tool for defining the sustainability problem. These consist of at least the economic, social, and environmental pillars, shown in Fig. 10.2 [5].

Fig. 10.1 The three Ps of sustainability: people, planet, profit [4]

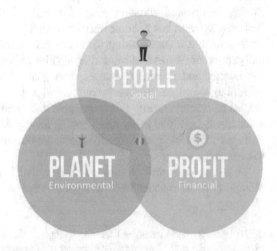

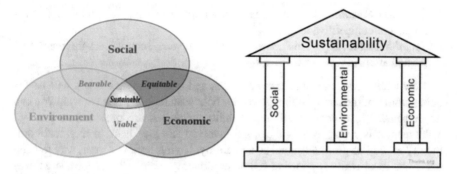

Fig. 10.2 The three pillars of sustainability [5]

Fig. 10.3 Sustainability in
the interaction of
environment, social
actions, and economics [7]

The four pillars of sustainability analysis are energy, efficiency, environment, and society. Sustainability analysis is multidisciplinary in nature. It requires approaches from different disciplines such as manufacturing engineering, environmental engineering, optimization, social science, and finance. Sustainability starts with green manufacturing and extends to industrial networks and then to the ecosystem. Sustainability of a system is its ability to survive and retain its functionality over time. A distinctly sustainable society is capable of surviving and prospering indefinitely [6].

Today, sustainability is seen in three dimensions: environmental, economic, and sociocultural, which are illustrated in Fig. 10.3 and explained as follows [7].

- *Environmental Sustainability*: The earth's resources and processes are connected with human societies. Environmental sustainability describes a possible way that human societies can sustainably develop by living within the system earth and using the resources of planet earth. It is focused on three protection goals: protection of human health, resources, and the ecosystem.

- *Economic Sustainability*: This addresses effective investments, finance, job creation, and competitiveness.
- *Social Sustainability:* This addresses equity, justice, security, employment, and participation.

New technology, new business practices, and new lifestyle models will be the cornerstones of the new sustainable world. Four sustainability tips for what you can do personally to support local sustainability initiatives are provided as follows. Volunteering is a way everyone can get involved and give back to the local community as well as promote social sustainability. Giving back is a great way to keep resources local while supporting fellow community members and local organizations. Communicate with your neighbors and community members about sustainability and social issues. Identify the challenges and opportunities that you could address to make positive changes in your community.

10.3   What Is Sustainable Manufacturing?

Sustainable manufacturing (SM) (or green manufacturing) can be defined as a method for manufacturing that minimizes waste and reduces the environmental impact. These goals will be obtained mainly by adopting practices that will influence the product design, process design, and operational principles. Therefore, sustainable manufacturing may be regarded as an approach that integrates product and process that will consider sustainability at all levels of the life cycle of manufacturing: product, process, and system. It promotes eliminating production and processing wastes through eco-efficient practices and encourages adopting new environmental technologies. The six major elements significantly affecting the sustainability of manufacturing processes are shown in Fig. 10.4 [8].

Sustainable manufacturing involves developing sustainable products with total life-cycle considerations. It is the creation of manufactured products using nonpolluting, natural resources conservation practices, supported by economically sound and safe manufacturing processes, practices, and systems. Such manufacturing practices are safe and are economically sound while simultaneously being societally beneficial.

Sustainable manufacturing requires that all manufacturers should aim for the following four activities that would help the environment across its entire supply chain [1]: (1) energy use reduction, (2) water use reduction, (3) emissions reduction, and (4) waste generation reduction. Sustainable manufacturing should integrate the sustainable activities at all levels of manufacturing—product, process, and system. This may involve the following 9R: reduce, reuse, recycle, recover, redesign, remanufacturing, repurpose, refurbish, and refuse. Industry 4.0 provides opportunities for the realization of sustainable manufacturing. Fig. 10.5 shows the key contributors to sustainable manufacturing [9].

Fig. 10.4 Six major elements affecting the sustainability of manufacturing processes [8]

Fig. 10.5 Key contributors to sustainable manufacturing [9]

Sustainable manufacturing is causing companies to implement new design and analysis procedures, energy reduction methods, material reduction efforts, and improved materials handling practices. Reducing consumption of energy, raw materials, water, and other resources in a factory to leverage greater efficiency and productivity is challenging and often starts with the basics of switching off lights and replacing the luminaires with LEDs. Fabric waste is turned into tiles and furniture.

Sustainable manufacturing has become the most important aspect to be considered by all manufacturing engineers otherwise known as production engineers, because it is an obligation to the world we live in. The three major principles sustainability leaders, actors, and implementers should keep in mind are reducing the resource utilization, using environment-friendly materials, and reducing all forms of waste while advocating for the reuse and recycling of materials.

The manufacturing industry seeks indicators to measure the sustainability of manufactured products and manufacturing processes. The main tool commonly used to implement SM is the life cycle assessment (LCA). It is a method used in assessing environmental impacts associated with all the stages of a product's life, from cradle to grave. It is an approach to examine fully the environmental impact of different activities performed by humans including the production of goods and services by corporations. LCA is mainly concerned with identifying the environmental impact of a given product or process at each stage of their life.

10.4 Sustainable Approaches

The issue of sustainability is becoming more and more central to the industry. Manufacturers engaged in sustainability activities include those of all sizes, ages, and sectors. The reasons companies are pursuing sustainability goals and initiatives include [10]:

- Increase operational efficiency by reducing costs and waste.
- Respond to or reach new customers and increase competitive advantage.
- Protect and strengthen brand and reputation and build public trust.
- Build long-term business viability and success.
- Respond to regulatory constraints and opportunities.

Ways that companies progress further on the path to sustainability include [10]:

1. Address sustainability in a coordinated, integrated, and formal manner, rather than in an ad hoc, unconnected, and informal manner.
2. Focus on increased competitiveness and revenues rather than primarily focusing on cost-cutting, risk reduction, and improved efficiency.
3. Use innovation, scenario planning, and strategic analysis to go beyond compliance.
4. Integrate sustainability across business functions.
5. Focus more on the long term.

6. Work collaboratively with external stakeholders.

Companies can build sustainable manufacturing for the new normal. To practice sustainable manufacturing in your company requires the following steps [11]:

1. Start practicing sustainable manufacturing with a plan.
2. Keep the benefits at the forefront of your focus.
3. Get certified in ISO standards that encourage environmental sustainability.
4. Analyze your production process and see what can be eliminated.
5. Encourage your entire company to be involved.

10.5   Sustainable Manufacturing Examples

Sustainable manufacturing is the creation of manufactured products using economically sound, nonpolluting processes which minimize negative environmental impacts while conserving energy and natural resources as well as increasing productivity and efficiency. It is relevant in different sectors like business, economics, environment, and society. Some examples of sustainable manufacturing include [12–14]:

- *Automotive Industry*: The automotive industry is regarded as a major economic force worldwide. It has undergone substantial transformations, which have reduced fuel consumption, minimized environmental impact, and improved safety. Sustainable manufacturing in the automotive industry is becoming an imperative strategy to drive profitable growth and provide value to customers. As the automotive industry continues to embrace and make sustainability its strategic imperative, it is important that automotive organizations reflect the newer 3P definition of sustainability: Pollution Prevention Pays [15]. In 2008, GM, a multinational corporation, approved three sustainability metrics: an energy use index, a water index, and a carbon emission index to calculate the company's performance and targets. Ford strove to eliminate waste from production. BMW was rated the world's most sustainable company in 2016.
- *Construction Industry*: The construction industry has its footprints on all human efforts to control, modify, and dominate nature and natural systems. There is a growing consensus that delivering a sustainable culture and environment starts with incorporating sustainability thoughts at the planning and design stages of an infrastructure construction project. Geotechnical engineering can significantly influence the sustainability of infrastructure development because of its early position in the construction process [16].
- *Furniture Manufacturing:* A furniture production company can integrate sustainability concepts in order to make a positive impact on the environment, society, and its own financial success. The principles of lean and green manufacturing

can deliver a significant, positive impact on multiple measures of operational performance of furniture production [17].

- *Pharmaceutical Manufacturing*: This relies on resources such as energy and water that can be expensive and generate large amounts of waste both toxic and benign. A robust focus on corporate sustainability can benefit pharmaceutical manufacturers. A sustainability strategy that focuses on managing risks associated with water use, energy consumption, and waste can prevent FDA warning letters [18].
- *Chemical Manufacturing*: Chemical manufacturing produces a large amount of wastes and also carbon emissions that harm the environment. Any process that reduces carbon emissions improves the sustainability of operations. Fossil fuels can be replaced with renewable energy sources in chemical manufacturing to facilitate sustainable production of high-quality chemicals [19].
- *Customized Manufacturing*: The production of personalized products using sustainable manufacturing systems and supply chains allows localized manufacturing and therefore a shortening of supply chains becoming more energy and resource efficient. Reshoring the apparel manufacturing to allow for faster, more customizable fashion would redefine the interaction between the manufacturer and the consumer. It is expected that this shift to more localized manufacturing will primarily affect goods manufacturers [20].
- *Remanufacturing*: This involves the processing of used products for restoration to their original condition. Remanufacturing and product recovery attracts significant attention due to environmental concerns, legislative requirements, consumer interests in green products, and market image of manufacturers.
- *Resource conservation*: The pressing needs of energy, water, and other resource conservation worldwide is a major engineering challenge. The most recent data on water use in the United States reported manufacturers consumed approximately 21 billion gallons per day from both municipal and self-supplied sources. As drivers such as population growth and climate change increase pressure on freshwater resources, both at the local and global level, manufacturers are seeking ways to incorporate more efficient and sustainable water use practices into their operations. This sustainable water use is driven from both an environmental perspective and a business perspective.
- *Sustainable Logistics*: Transportation is an important part of the manufacturing supply chains. There is a need for a sustainable means for shipping goods from origin A to destination B. Sustainable shipping works toward reducing transportation emissions footprint that can be harmful to the environment [21].
- *Regulatory Compliance:* A sustainability strategy that focuses on managing risks associated with water use, energy consumption, and waste can prevent FDA warning letters.

10.6 Benefits

Environmental responsibility has become an integral part of the way products are manufactured, marketed, and purchased. The benefits of sustainable manufacturing are almost infinite. An increasing number of manufacturers are realizing substantial benefits from sustainable business practices. Some forward-**thinking** organizations have added vice presidents of sustainability to their leadership teams. Designing products to be environmentally benign can contribute to their successful introduction and maintenance. Other benefits of sustainable manufacturing include the following [22, 23]:

- *Increased Sales:* Sustainable manufacturing will make your business more attractive and marketable. Sustainability should be particularly profitable for manufacturers.
- *Energy Efficiency:* Companies of all sizes are pursuing sustainability by improving efficiency and reducing their energy consumption. When manufacturing focuses on efficiency, energy use invariably decreases.
- *Save Energy Costs:* Replacing incandescent bulbs and fluorescent tubes with LED lighting can reduce electricity usage because LED lighting consumes a smaller amount of electricity. Converting to a renewable energy source, such as wind, geothermal, or solar, can stabilize energy cost with a much longer payback period.
- *Incentives:* There are government incentives, tax credits, grants, and utility company rebates for businesses that support sustainable practices.
- *Workplace Morale:* Implementing green practices typically spurs collaboration and teamwork.
- *Recycling of Waste:* Implementing sustainable practices in manufacturing reduces water and energy usage, minimizes waste, and decreases hazardous emissions. Production waste is collected by the manufacturing company or by a specialized recycling company and returned for recycling.
- *Environmental Health and Safety:* Sustainable manufacturing enhances the safety of the products, employees, and community. Safety practices include developing and enforcing employee safety procedures. Such practices can return your investment by providing an accident-free workplace and demonstrating to your employees that their health and safety are important.
- *Quality Improvement:* Quality improvement programs offer your best employee-driven opportunities for enhanced teamwork while registering satisfied customers, sales growth, and efficient operation.
- Reduce costs, reduce waste, and increase operational efficiency.
- Increase competitive advantage.
- Enhance brand name recognition and reputation while simultaneously cultivating public trust and loyalty.
- Contribute to long-term business visibility, viability, and success.
- Comply with regulatory constraints.

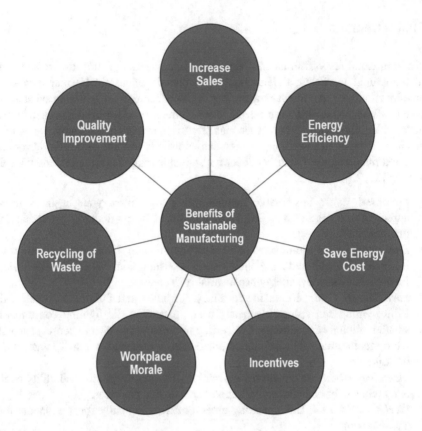

Fig. 10.6 Some benefits of sustainable manufacturing

Some of these benefits are illustrated in Fig. 10.6. Other benefits of sustainable manufacturing include improved morale, improved brand value, lowered regulatory concerns, increased market opportunities, improved product performance, and decreased liabilities. SM considers the cost of compliance to the environmental guidance prescribed in relation to the value to society and the adverse impact imposed on society and the manufacturing industry, such as job flight to foreign low wage jurisdictions, caused by the prescribed sustainability regulation(s).

10.7   Challenges

The major challenges faced by the manufacturing industry in its pursuit of sustainability goals are as follows [24]:

1. The manufacturing industry is facing the challenge of measuring sustainability performance in a product's life cycle. Developing metrics for sustainable manufacturing is critical to enable manufacturing companies to quantitatively measure the sustainability performance in specific manufacturing processes.

2. Industry is unable to measure economic, social, and environmental impacts and costs of their products accurately during the entire life cycle and across their supply chain.
3. Full life cycle analysis or assessment (LCA) of products requires new methods to analyze, integrate, and aggregate information across hierarchical levels, organizational entities, and supply chain participants. Existing methods of aggregation do not take into account sustainability issues.
4. Cost is one of the main reasons manufacturers stay away from sustainable practices.
5. Industry lacks neutral and trusted standards and programs to demonstrate, deploy, and accredit new sustainable manufacturing practices, guidelines, and methods.
6. Regulations need to be supported and informed by industry standards. These regulations/standards should be harmonized.
7. Current manufacturing modeling and assessment criteria require intensive revisions and upgrades to keep up with these new challenges.
8. The production and delivery of personalized goods and services using sustainable manufacturing systems and processes present a major challenge.
9. The word "sustainability" has been overused and abused.
10. Consumer education is needed and in some cases critical, as some customers are not sure what packaging can and cannot be recycled.
11. There are too many metrics; these can be condensed and grouped, predicated on consolidation and harmonization.

10.8   Global Adoption of Sustainable Manufacturing

With Goal 12 of the UN Sustainable Development Goals encouraging responsible production, sustainable practices have been gaining traction worldwide. An increasing number of organizations are treating "sustainability" as an important part of their strategy to increase growth and global competitiveness. Manufacturers around the world are making the move toward adopting sustainability. We now consider how different nations are integrating sustainability into their manufacturing and design.

- *United States:* Fig. 10.7 shows the environmental impact of the US manufacturing industry [25]. The US Environmental Protection Agency (EPA) describes sustainable manufacturing as "the creation of manufactured products through economically sound processes that minimize negative environmental impacts while conserving energy and natural resources. Sustainable manufacturing also enhances employee, community, and product as well as process safety." While the manufacturing industry is a driving force of the American economy, it also has a sizeable carbon footprint. In recent times, there has been renewed interest in US manufacturing and endeavors launched for sustainable manufacturing ini-

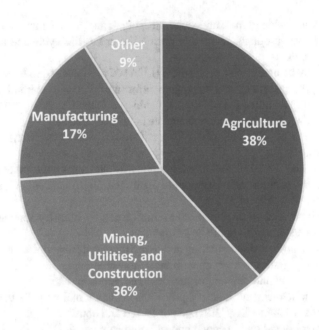

Fig. 10.7 Environmental impact of the US manufacturing industry (modified) [25]

tiatives. For example. The Kentucky Sustainable Manufacturing Initiative (KSMI) is designed to assist manufacturers with learning how to integrate sustainability into their daily manufacturing operations.

- *United Kingdom:* It has been recommended that food and drink manufacturers become key agents of change. In response to this, Coca-Cola Enterprises (CCE) has released the final findings, entitled Sustainable Manufacturing for the Future. It has launched a £56 m operational investment plan, accelerating its journey toward sustainability in Great Britain. CCE shares its vision for sustainable manufacturing, offering a picture of what the "factory of the future" may look like in Great Britain by 2050 [26].

- *Canada:* According to the government of Canada, sustainable manufacturing promotes minimizing or eliminating production and processing wastes through eco-efficient practices and encourages adopting new environmental technologies. Although the manufacturing sector (including primary metal, paper, chemical, and petroleum and coal) significantly contributes greatly to the Canadian economy, the sector has contributed to the inefficient utilization of energy and increased pollution, posing a threat to the environment. The Canadian government is therefore encouraging manufacturing organizations to help fulfill its commitment to reduce the nation's greenhouse gas (GHG) emissions by 20 percent in the year 2020. Adopting sustainable manufacturing practices can help manufacturing companies reduce their GHG emissions, enhance their brand image, gain a competitive edge, and build trust among the investors, regulators, and customers [27].

- *China:* Chinese manufacturing industry is in a transitional period. Although China is making great developments industrially, there is still a lot of ground to cover. China has launched the Made in China 2025 initiative to bring the Chinese economy to the cutting edge and create a sustainable manufacturing base. By employing AI, China is modernizing their manufacturing and economy. China is also investing in the development of sustainable manufacturing. It will become one of the fastest growing sustainable manufacturing sectors [28].
- *Norway:* Norwegian society is dependent on sustainable value creation. Manufacturing products and services based on the natural factors of our environment should be given priority. A reindustrialization process is also needed so that manufacturing previously outsourced to low-cost countries can return to Norway. The Government's Industry White Paper of March 2017 highlights the need using new materials, digitizing processes, restructuring within sustainable limits, and emphasizing the importance of research and innovation. To attract Norwegian and international players to invest in Norway, it will be necessary to demonstrate that the activity is sustainable [29].
- *India:* A survey of 198 Indian SMEs have identified the following aspects of sustainable manufacturing: "The final quantitative benefits of green manufacturing in order of their decreased ranking are improved morale, improved brand value, lowered regulatory concerns, increased market opportunities, improved product performance, and decreased liabilities. The government has earmarked funds for improving the current infrastructure. Several measures have been taken toward the sensitization on sustainable agricultural methods including optimum use of water and saving as much water as possible for irrigation [30].

10.9 Conclusion

Protecting our planet is becoming a priority for everyone. Today, there is a growing awareness of environmental stewardship and sustainability. Sustainability has become an increasingly important requirement for economic activities. It has been applied to many fields such as manufacturing, design, engineering, and environmental stewardship. The implementation of sustainable systems is an essential requirement in modern manufacturing.

A growing number of manufacturers are treating "sustainability" as an important objective in their strategy and operations in order to increase growth and global competitiveness. Sustainable manufacturing, with promising environmental and social benefits, is the wave of the future for manufacturing. It needs to be integral to every aspect of a manufacturer's operations. More information about sustainable manufacturing can be found in the books in [31–53] and the related journals:

- *Sustainability*
- *Manufacturing*
- *International Journal of Sustainable Manufacturing*

- *International Journal of Sustainable Engineering*
- *International Journal of Precision Engineering and Manufacturing-Green Technology*
- *CIRP Journal of Manufacturing Science and Technology*
- *Smart and Sustainable Manufacturing Systems*

References

1. N. Posinasetti, Sustainability sustainable manufacturing: principles, applications and directions, May 2018, https://www.industr.com/en/sustainable-manufacturing-principles-applications-and-directions-2333598
2. M. Abraham, Sustainable engineering: an initiative for chemical engineers. Environ Prog **23**(4), 261–263 (December 2004)
3. M.N.O. Sadiku, S.R. Nelatury, S.M. Musa, Green engineering: A primer. J Scientif Eng Res **5**(7), 20–23 (2018)
4. E. Avramenko, Contribute to a better world = Be a successful company. Sustainability in a shared-economy startup, February 2018, https://medium.com/@Anaiska/contribute-to-a-better-world-be-a-successful-company-sustainability-in-a-shared-economy-startup-50cc72f2e37b
5. The three pillars of sustainability, https://www.thwink.org/sustain/glossary/ThreePillarsOfSustainability.htm
6. M.N.O. Sadiku, O.D. Olaleye, S.M. Musa, Sustainable engineering: an introduction. Int J Adv Scientif Res Eng **5**(6), 70–74 (June 2019)
7. Athena, Sustainable development, http://macaulay.cuny.edu/eportfolios/akurry/2011/12/21/sustainable-development/
8. K. R. Haapalam et al., Review of engineering research in sustainable manufacturing, *Proceedings of the ASME* 2011 International manufacturing science and engineering conference, Corvallis, Oregon, USA, June 13-17, 2011
9. H.A. Kishawy, Sustainable manufacturing and design: concepts, practices and needs. Sustainability **4**(2), 154–174 (2012)
10. EPA, Sustainable manufacturing, https://www.epa.gov/sustainability/sustainable-manufacturing#:~:text=Sustainable%20manufacturing%20is%20the%20creation,employee%2C%20community%20and%20product%20safety
11. 5 Tips to practice sustainable manufacturing, December 2020, https://industrytoday.com/5-tips-to-practice-sustainable-manufacturing/
12. Sustainable engineering products and manufacturing technologies, 2019, 13th *Proceedings of International Symposium on Process Systems Engineering* (PSE 2018), 2019
13. US Department of Energy, Chapter 6: Innovating clean energy technologies in advanced manufacturing, July 2015, https://www.energy.gov/downloads/chapter-6-innovating-clean-energy-technologies-advanced-manufacturing
14. F. Sanger, Sustainability tip: think local first how giving back to your community relates to sustainability, October 2018, https://blog.walkingmountains.org/sustainability/sustainability-tip-think-local-first-how-giving-back-to-your-community-relates-to-sustainability
15. G. Appu, Sustainable manufacturing in automotive industry and how it can be game-changer, March 30, 2021. https://www.financialexpress.com/auto/industry/sustainable-manufacturing-automotive-industry-game-changer-maruti-hyundai-tvs-hero-ashok-leyland/2222895/
16. D. Basu, A. Misra, A.J. Puppala, Sustainability and geotechnical engineering: perspectives and review. Can Geotech J **52**, 96–113 (2015)
17. G. Miller, J. Pawloski, C.R. Standridge, A case study of lean, sustainable manufacturing. J Indust Eng Manage **3**(1), 11–32 (2010)

18. S. Fotheringham, Reining in consumption for sustainable manufacturing, March/April 2018., https://ispe.org/pharmaceutical-engineering/march-april-2018/reining-consumption-sustainable-manufacturing
19. Discover 5 top sustainable manufacturing solutions, https://www.startus-insights.com/innovators-guide/discover-5-top-sustainable-manufacturing-solutions/
20. How accurate's sustainable manufacturing can benefit your project, June 2019, https://www.accurateperforating.com/blog/how-accurate%E2%80%99s-sustainable-manufacturing-can-benefit-your-project
21. E. Raw, 3 Sustainable manufacturing trends for 2020 and beyond, https://www.reliableplant.com/Read/31850/sustainable-manufacturing
22. K. McAslan, Reasons to embrace sustainable manufacturing: the new industry standard, May 2018, https://www.cobizmag.com/reasons-to-embrace-sustainable-manufacturing/
23. B. Frahm, Building a sustainable manufacturing enterprise, https://www.thefabricator.com/thefabricator/article/shopmanagement/building-a-sustainable-manufacturing-enterprise
24. S. Rachuri et al. (eds.), Sustainable manufacturing: Metrics, standards, and infrastructure - NIST Workshop Report, April 2010, https://www.nist.gov/publications/sustainable-manufacturing-metrics-standards-and-infrastructure-nist-workshop-report
25. K. C. Morris, Sustainable manufacturing is smart manufacturing, October 2, 2020, https://www.nist.gov/blogs/taking-measure/sustainable-manufacturing-smart-manufacturing
26. J. Williamson, Coca-Cola Enterprises launches vision for sustainable manufacturing, March 2016, https://www.themanufacturer.com/articles/coca-cola-enterprises-launches-vision-for-sustainable-manufacturing/#:~:text=Coca%2DCola%20Enterprises%20launches%20vision%20for%20sustainable%20manufacturing,-Posted%20on%20 22&text=Coca%2DCola%20Enterprises%20(CCE),Sustainable%20Manufacturing%20 for%20the%20Future.&text=These%20themes%20set%20the%20agenda%20for%20the%20 partnership's%20next%20phase%20of%20research
27. 5 Considerations for employing sustainable manufacturing practices, https://www.canadianmetalworking.com/canadianmetalworking/article/management/5-considerations-for-employing-sustainable-manufacturing-practices
28. Sustainability in manufacturing: Made in China 2025 and the BRI, December 2018, https://et2c.com/sustainability-in-manufacturing/
29. Competitive and sustainable manufacturing, https://www.ntnu.edu/iv/competitive-and-sustainable-manufacturing
30. How Indian Fintechs played a pivotal role in enabling Atma Nirbhar Bharat under various domains, http://bwsmartcities.businessworld.in/article/How-Indian-Fintechs-Played-A-Pivotal-Role-In-Enabling-Atma-Nirbhar-Bharat-Under-Various-Domains-/30-12-2020-359724/
31. G. Seliger, *Sustainability in manufacturing: recovery of resources in product and material cycles* (Springer Science & Business Media, 2007)
32. G. Seliger (ed.), *Sustainable Manufacturing: Shaping Global Value Creation* (Springer, 2012)
33. J. Kauffman, K. Mo, *Lee, handbook of sustainable engineering* (Springer Verlag, 2013)
34. A. Pampanelli, N. Trivedi, P. Found, *The Green Factory: Creating Lean and Sustainable Manufacturing* (Productivity Press, 2016)
35. S. Vinodh, *Sustainable manufacturing: concepts, tools, methods and case studies* (CRC Press, Boca Raton, FL, 2020)
36. J.P. Davim (ed.), *Sustainable Manufacturing* (Wiley-ISTE, 2010)
37. K. Salonitis, K. Gupta (eds.), *Sustainable Manufacturing* (Elsevier, 2021)
38. G. Seliger, Sustainable manufacturing: shaping global value creation.. Springer Science & Business Media, 2012
39. W. Li, S. Wang, *Sustainable manufacturing and remanufacturing management: process planning, Optimization and Applications* (Springer, 2018)
40. M. Singh, T. Ohji, R. Asthana, *Green and sustainable manufacturing of advanced material* (Elsevier, 2015)

41. G. Seliger, M.M.K. Khraisheh, I.S. Jawahir, *Advances in sustainable manufacturing* (Springer Science & Business Media, 2011)
42. K. Gupta, *Advanced manufacturing technologies: modern machining, Advanced Joining, Sustainable Manufacturing* (Springer, 2017)
43. M.J. Franchetti, B. Elahi, S. Ghose, *Value creation through sustainable manufacturing* (Industrial Press, 2016)
44. S. Roberts, *Sustainable manufacturing?: the case of South Africa and Ekurhuleni* (Juta and Company Ltd, 2006)
45. A.N. Nambiar, A.H. Sabuwala, *Sustainable manufacturing* (CRC Press, Boca Raton, FL, 2014)
46. I. Garbie, *Sustainability in Manufacturing Enterprises: Concepts, Analyses and Assessments for Industry 4.0* (Springer, 2016)
47. R. Dubey, Strategic Management of Sustainable Manufacturing Operations.. IGI Global, 2016
48. D. Rickerby (ed.), *Nanotechnology for sustainable manufacturing* (CRC Press, Boca Raton, FL, 2014)
49. N.K. Jha, *Green design and manufacturing for sustainability* (CRC Press, Boca Raton, FL, 2015)
50. K. Kumar, D. Zindani, in *Sustainable Engineering Products and Manufacturing Technologies*, ed. by J.P. Davim, (Academic Press, 2019)
51. K. Kumar, D. Zindani, in *Sustainable Manufacturing and Design*, ed. by J.P. Davim, (Woodhead Publishing, 2021)
52. K. Jayakrishna et al. (eds.), *Sustainable manufacturing for industry 4.0: an augmented approach* (CRC Press, Boca Raton, FL, 2020)
53. R. Stark, G. Seliger, J. Bonvoisin (eds.), *Sustainable Manufacturing: Challenges, Solutions and Implementation Perspectives (Sustainable Production, Life Cycle Engineering and Management)* (Springer, 2017)

Chapter 11
Lean Manufacturing

Lean manufacturing is at the core of our strategy to support our growth. You have to take waste out of everything. We only want to do things that our customers are willing to pay for.

– David Buck.

11.1 Introduction

Due to global competition and the competitive awareness of customers, manufacturers all over the world are under pressure to reduce their production costs, eliminate waste, increase productivity, and improve quality while remaining profitable. To survive in the prevailing global competition, a manufacturing company must eliminate production waste. One approach that is widely applied in eliminating waste is lean manufacturing. Lean manufacturing (LM) is all about reducing or eliminating waste in production in all its forms. It is "manufacturing without waste." It is widely perceived by the industry as an answer to their requirement for waste reduction [1, 2].

Lean manufacturing (LM) is all about doing more with less by applying "lean thinking." Lean thinking is believing that there is a simpler, better, easier way to complete our work. LM involves conscious, never-ending efforts to completely eliminate or reduce waste. Its goal is to satisfy the customer with the exact product, quality, and quantity delivered to the right location at the right time and price in the shortest amount of time.

Lean principles are derived from methodologies applied by the Japanese manufacturing industry in the 1990s. It is a systematic method that focuses on eliminating or reducing waste within manufacturing systems without sacrificing productivity. It is the application of lean practices, principles, and techniques to manufacture products as well as deliver services. It has become a popular concept in the developed world as well as in the developing world. It is expressed in the famous adage that says, "a penny saved is a penny earned."

This chapter provides an introduction to lean manufacturing, its philosophy and applications. It begins by describing different types of waste. It explains what lean manufacturing is all about. It covers lean manufacturing principles. It mentions the relationship between lean manufacturing and other manufacturing concepts such as

sigma six, green manufacturing, and Industry 4.0. It provides some reasons companies are implementing lean manufacturing. It gives some applications of lean manufacturing. It highlights the benefits and challenges of lean manufacturing. It
addresses the global adoption of lean manufacturing. The last section concludes
with comments.

11.2   Types of Waste

Waste is any process or activity without value such as overproduction, defects, non-
value-added motion, redundancy of processing, idleness, and waiting. It is anything
besides the essential resources of people, machines, and materials that are needed to
add value to the product. Any action which does not directly enhance the product's
value can be considered as waste. There are seven forms of waste that are now generally accepted [3]: overproduction, inventory, waiting, quality, waiting, unnecessary transport, unnecessary movement, and overprocessing. These are illustrated in
Fig.11.1 [4] and explained as follows [5].

- *Overproduction Waste:* This is producing more than the customer demands.
  There are two types of overproduction: (1) quantitative, making more products
  than needed, and (2) early, making products before needed.
- *Inventory Waste*: This involves keeping more than the minimum stock of raw
  materials, parts, and finished goods necessary. To prevent this, list only what
  you need.
- *Waiting Waste:* Spending time waiting is called spending for no reason. This
  includes waiting for material, labor, information, equipment, etc. It also includes

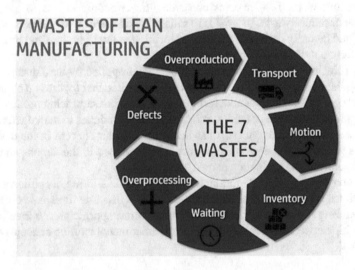

Fig. 11.1 The seven wastes of lean manufacturing [4]

operators standing idle while machines or equipment fails. When big mistakes are made, resources are wasted.

- *Defect Waste*: This entails the time and effort spent correcting and inspecting rework and scrap. When something is not working properly, there is defect waste. Learn to identify common mistakes and make everyone aware of them so that the mistakes do not occur consistently.
- *Transportation Waste:* Excessive movements and handling can cause damages and can lead to quality deterioration. Transportation waste can be a catalyst for other wastes such as waiting.
- *Motion Waste:* If something is moved when it is not really necessary, that is regarded as motion waste. Any motion that an employee has to perform that does not add value to the product is considered waste.
- *Overprocessing*: This consists of steps that add no value to the product but cost the outlay of resources. It includes any activities performed on the products that are not required or expected by the customer. Overprocessing costs time, materials, machinery, and money.
- *Underutilized Talent.* This refers to not utilizing people to their fullest potential including time, talent, skill, and even ideas. To make sure your workers are reaching their full potential, invest in quality training programs that build efficiency within your teams.

The acronym TIIMWOOD can be used to remember the eight types of waste. It stands for: **T**ransport, **I**dleness, **I**nventory, **M**otion, **W**aiting, **O**verproduction, **O**verprocessing, **D**efect. Thus, lean manufacturing is a generic process that aims at getting the right things to the right place, at the right time, and in the right quantity to achieve zero waiting time, zero inventory, reduced batch sizes, and reduced process times.

11.3 Concept of Lean Manufacturing

Lean manufacturing (also known as lean production) is regarded as a methodology to improve productivity and decrease costs in manufacturing organizations. It addresses one of the worst things that can happen to any company: waste. It is regarded as a declared war against the waste of both manufacturing inefficiencies and the underutilization of people. It is also concerned with the company's goals of improving quality, reducing inventory, and becoming more competitive. Lean can mean "less" in terms of less waste, less design time, less cost, fewer organizational layers, and fewer suppliers per customer. The major goal of lean manufacturing is total elimination of waste in human effort, inventory, and time. Other goals of lean manufacturing are to improve the quality of products, reduce production times, and reduce total costs [6].

Lean manufacturing is American name for the Toyota Production System which was introduced by Toyota in the 1990s for the minimization of waste and inefficiency in its manufacturing operations. Today, hundreds of manufacturing companies around the world embrace the principle of lean manufacturing to reduce waste and improve their overall operations and enhance their profitability. For an American manufacturing company, being lean manufacturing is crucial, for competing with lower-cost countries. Lean aims to enhance productivity by simplifying the operational structure and the work environment [7].

To remain competitive, manufacturing organizations have to be in a constant mode of continuous improvement. Lean manufacturing continually focuses on management efforts to improve processes to reduce or eliminate any wastes (deviations from standard) through employee involvement in continuous improvement of products, processes, and standards. As waste is eliminated, production time and cost are reduced, and organizational profits soar.

11.4 Lean Manufacturing Principles

Womack and Jones, in their book *Lean Thinking: Banish Waste and Create Wealth in Your Corporation*, provide the following five principles of lean manufacturing [8–10].

- *Value:* The critical starting point for lean thinking is value. The value the customer assigns to the product and services is what determines what the customer will pay. The manufacturer focuses on eliminating waste so that it can deliver the value the customer expects at the highest level of profitability.
- *Value Stream:* Once the value (end goal) is determined, the next step is mapping the "value stream." This refers to the totality of the product's entire life cycle from the raw materials to eventual disposal. All businesses must clarify the "value stream," the nexus of actions to bring the product through problems solving, information management, and physical transformation tasks. Steps, materials, features, and movements that do not add value are eliminated.
- *Flow:* After the waste has been eliminated from the value stream, the next step is to be sure that the remaining steps flow smoothly with no interruptions, delays, or bottlenecks. Flow traces the product across departments. Understanding flow is essential to the elimination of waste. The lean manufacturing principle of flow is about creating a value chain with no interruption in the production process.
- *Pull:* The lean principle of pull helps ensure flow by making sure that nothing is made ahead of time. The pull approach dictates that nothing is made until the customer orders it. This means that the customer can "pull" the product from the manufacturer as needed. This way products do not need to be built in advance and stored.
- *Perfection:* This involves making lean thinking and process improvement part of the corporate culture. Lean practitioners strive to achieve nothing short of perfection. The relentless pursuit of perfection is what drives users of the approach to dig deeper, measure more, and change more often than their competitors. Every

Fig. 11.2 Five key lean principles [6]

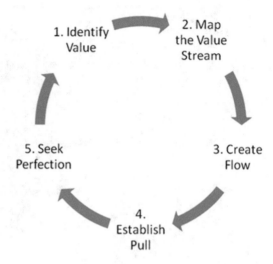

employee should be involved in implementing lean. A collective mindset for continuous and incremental improvement is essential to reach the company's goal of perfection.

These principles of the lean approach are illustrated in Fig. 11.2 [6]. They have transformed countless corporations, giving them a leg up on the competition and a clear path to both profitability and delighted customers. Whatever we do under the lean umbrella should always answer these principles. For lean principles to be effective, lean must be applied throughout an organization.

Other principles of lean manufacturing constitute a 5S strategy, which should be an integral part of the organization's work culture. The principles are meant to increase efficiency by curtailing the wastage of effort, time, and money. Figure 11.3 shows the five pillars of the 5S system [11]. Here's what each "S" in 5S stands for [12]:

1. *Sort:* The participants or employees sort through their tools and equipment, separate that which is needed, and transfer them to their designated storage locations.
2. *Straighten:* Once the first step is complete, the employees can further improve efficiency by arranging the items in a way that promotes minimal motion, transportation, and waiting time. Tools and other items should always be arranged skillfully in a way that employees can easily locate them.
3. *Shine:* The aim of shine is to have a clean workplace to ensure that the workplace is free of dirt. It consists of cleaning and inspecting regularly using tools to keep them in good condition. Shining is cleaning regularly so other lean standards can be upheld and so defects are not missed.
4. *Standardize:* The main objective of the standardization step is to provide standard operating procedures for 5S implementation. This could be a standard method of reviewing and assessing the 5S strategy.

Fig. 11.3 Five pillars of
the 5S system [11]

5. *Sustain*: This ensures the sustainability of the 5S strategy by creating constant reminders for the workers to implement 5S in the best possible manner.
6. *Safety:* Modern standards are adding safety as an additional S, creating a 6S strategy. Neglecting safety can cause loss of life, harm to the environment, and loss of reputation of the company.

It is worth mentioning that lean manufacturing is a Japanese method focused on 3 Ms.: *muda,* the Japanese word for waste; *mura,* the Japanese word for inconsistency; and *muri,* the Japanese word for unreasonableness.

11.5   Relationship of Lean and Other Concepts

Lean manufacturing is the optimal way of producing goods by minimizing waste.

It is a philosophy that is based on creating smooth flow, minimizing waste, and respect for the worker.

It may be expedient to take a moment to see the relationship between lean and other manufacturing concepts such as Six Sigma, green manufacturing, cellular manufacturing, and Industry 4.0.

- *Six Sigma*: This was founded by Motorola in late 1980s. Six Sigma is the most popular quality movements in the industry. Six Sigma is a method that relies on a collaborative team effort to improve performance, remove waste, and reduce variation. The 3 Cs in Six Sigma are (1) change, changing society; (2) customer, power is shifted to customer; and (3) competition, competition in quality and productivity. Lean Six Sigma is a synergized managerial concept that combines lean and Six Sigma. It has been recommended as the integration of lean and Six

Sigma so that it becomes a much more powerful integrated tool. The process is used to eliminate defects in a process. It is a powerful tool to improve the efficiency and effectiveness of a business [13, 14]. Some organizations have combined the principles of lean manufacturing and Six Sigma into one program known as *Lean Six Sigma.*

- *Green Manufacturing*: This focuses on minimizing environmental impact of manufacturing processes and products. It refers to modern manufacturing that makes products without pollution. It addresses a wide range of environmental and sustainability issues including resource selection, transportation, manufacturing process, and pollution. Lean manufacturing includes several techniques to eliminate waste and improve the manufacturing system. Green manufacturing systems reduce waste just as lean manufacturing. Although not every lean practice is correlated with every environmental indicator, lean manufacturing is a good example of what is known as a sustainable competitive advantage.
- *Cellular Manufacturing:* This is a particular model for workplace design that is an important part of lean manufacturing. Successful implementation of manufacturing cells requires addressing product and/or process selection, design, operation, and control issue. The most important benefits of cellular manufacturing are achieved when manufacturing cells are designed, controlled, and operated using just-in-time (JIT) [15].
- *Industry 4.0:* This is the current gradual industrial transformation with automation, cloud computing, cyber-physical systems, robots, and industrial IoT to realize smart industry and manufacturing. It is the new manufacturing objectives with the aim of achieving nearly zero-defects production in the manufacturing industry [9]. Lean systems have an entirely different focus. Lean may be regarded as a combined set of principles, practices, and techniques with an aim to improve quality, cost, and customer satisfaction by eliminating three main sources of loss: variability, waste, and inflexibility [16].

11.6   Why Are Companies Using Lean Manufacturing?

Some typical reasons companies are using lean manufacturing are provided as follows.

- "Lean thinking" was adopted by the automobile industry business processes and operations.
- Lean principles help workers in the food industry to make improvements in performance.
- In plastic bags industry, lean manufacturing has caused higher productivity and identified the major types/sources of waste.
- Lean manufacturing principles have been implemented in a cement industry, which is a process industry, different from discrete manufacturing.

- Implementing lean in printing companies may dramatically enhance performance, save time, improve quality, lower costs, reduce inventory, and lead times.
- Lean principles can be applied in meal production and can result in increased production efficiency and systematic improvement of product quality.
- Application of lean manufacturing principle enabled Intel to meet a 46% increase in the number of training collateral with a 36% reduction in training resources.
- Lean is useful in hospitals and has made possible, limited impact on the overall healthcare performance.

11.7   Applications

Although lean principles have their roots in the manufacturing sector, they have been applied successfully in other industries. The principles can be applied to any business or production process in any industry. As typically illustrated in Fig. 11.4 [17], lean principles have been successfully applied to various sectors including automotive industry, education, electronics, healthcare industry, software development, service industry, construction industry, food industry, sewing and textile industries, call centers, supply chain, information technology, printing industry, hotel, tourism, transport, materials handling, etc.

- *Automotive Industry*: Lean manufacturing has revolutionized manufacturing starting in the automotive industry. The lean manufacturing concept was first coined in The International Motor Vehicle Program, attended by automotive

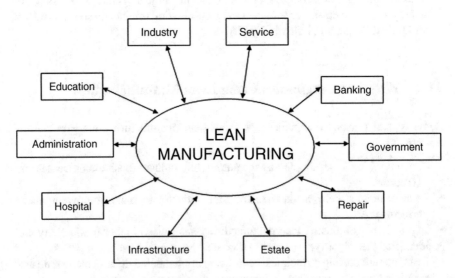

Fig. 11.4 Some applications of lean principles [17]

manufacturers in North America, Japan, and Europe. The lean manufacturing was launched as a concept describing best practice within the automotive industry but has gradually evolved to other industries [18, 19].

- *Plastic Industry:* Lean manufacturing is concerned with the implementation of several tools and methodologies that aim for the continuous elimination of wastes throughout the manufacturing process. It focuses on the company's goals of improving quality, reducing inventory, and becoming more competitive. Lean manufacturing has been implemented in the plastic bag industry. It has achieved lower process cycle time, lower manufacturing lead time, and higher productivity [20].

- *Machine Tool Industry*: This industry often requires quick delivery since machine tool companies are selling their products to all types of industries. These companies need machine tools as soon as possible to deliver their product. If a job shop receives some business that requires machine tools, then the customer will immediately need to acquire that specific machine, and quick delivery is very important. Manufacturers, such as Mazak Corporation in Florence, KY, are facing a difficult economic environment, as machine-tool technology supplier. Mazak has used lean manufacturing principles to develop a new production system called "modular assembly" in 2002 [21, 22].

- *Printing Industry:* The newspaper industry is facing several challenges, and its development is continuously changing. It is crucial for newspapers to analyze the trends in order to survive in the challenging market. The industry was still profitable by the end of 2008, but the recession has threatened the survival of newspapers. Implementing lean in printing companies is an excellent starting point in identifying and removing bottlenecks and constraints. It may dramatically enhance performance, save time, improve quality, lower costs, and reduce inventory [23].

- *Sewing Industry:* The sewing industry is an essential component of the textile industry. Sewing requires using sewing machines to assemble fabric pieces and attach various accessories like elastics, buttons, and labels. Sewing is considered a labor-intensive industry. Widespread implementation of lean manufacturing is considered recent to textile companies. However, lean manufacturing implementation in the sewing industry is challenging. To implement lean manufacturing tools in the sewing process, process measures must be defined, and the cost of processing must be quantified [24].

- *Lean Management*: This is a common application of lean principles. It encourages shared responsibility and shared leadership. The main objective of lean management is creating value to the customer by optimizing resources. The two main pillars of the lean methodology are respect for people and continuous improvements, ensuring that every employee is involved in the process of improving [25].

- *Quality:* Quality is very important in manufacturing. Customer demands for high-quality goods. Quality in manufacturing is more related to process control.

The risk of product-liability litigation against businesses has compelled companies to seek ways to sustain quality assurance in their products and services. With the rapid increase in the number of new product introductions, quality may be compromised. To stay competitive, companies must meet customers' changing wants and needs. Some companies are already ingraining quality into every step of their production process. The goal of lean manufacturing is producing quality products, reducing waste, and maximizing material flow. Lean manufacturing is one way to improve the quality of a product and service. Quality assurance may be regarded as assuring product quality in order to increase the customer's satisfaction and to reduce the total cost.

- *Product Safety:* This is a basic requirement from customers. Manufacturers pay substantial efforts to predict potential safety hazards on the prototypes so as to improve the product safety in new product development. Lean manufacturing can assist manufacturer to enhance the potential product safety issues in predesign and post-design stages of product development processes [26].
- *Hospital:* Lean is often applied in hospitals, but the impact tends to be limited. Lean tends to be applied in secondary and support functions with a logistic character and therefore has had a limited impact on the overall healthcare performance. Lean is useful for hospitals, but the lean concept as well as the implementation of lean in hospitals has often been described as a straightforward process. Lean provides opportunities for making significant improvements within all areas of a hospital [27].

Other lean manufacturing applications include aerospace, textile industry, food industry, material handling, and construction. Each of these applications may present its own challenges. The success of implementation of any particular application largely depends upon organizational characteristics. Lean manufacturing is a multidimensional approach, and not all organizations should implement the same set of practices [28].

11.8   Benefits

Lessons learned from lean principles can be universally applied to any business. The benefits of lean principles are evident in many factories worldwide. Every company has a tremendous opportunity to improve, using lean principles. First-pass correct output, reduced manufacturing lead time, and increased productivity are the three main drivers of lean implementation. The operational metrics for LM are on target and business-centric on all accounts: high productivity, reduced lead time, improved first-pass correct output, reduced inventory, and space requirement [29]. Other benefits include [30, 31]:

- *Reduce Waste:* Waste is bad for costs and resources. The lean manufacturing process helps companies reduce or eliminate the excess waste produced during production. Lean provides a way to do more and more with less and less. By focusing on eliminating waste, a company can reduce its negative impact on the global environment.
- *Reduce Costs:* Money is saved when a company is not wasting time, resources, and personnel on unnecessary activities. Overproduction adds to the cost for storage.
- *Reduce Time:* Time is money, as the adage goes, and wasting time is therefore wasting money. Reducing the time it takes to start and finish a project improves efficiencies. The less time that is required to complete a task, the leaner the workforce.
- *Safer Working Environment*: Safety and ergonomics are incorporated in lean manufacturing. Less inventory means less clutter, which means less accidents and better visibility. It also means a lesser chaotic environment and more free space.
- *Improve Employee Morale*: Constant feedback from employees includes and empowers them in the decision-making process. This boosts employee morale. Efforts toward less wasteful production can also make your employees' work more productive and improve employee morale.
- *Improve Quality:* Lean manufacturing utilizes fewer resources and produces high levels of product quality and service. Since there is less focus on excessive production, there is more of an opportunity to focus on quality.
- *Better Customer Service:* By reducing various types of waste in your company, you can get products to your customers, faster, and meet their needs more easily.
- *Better for the Environment*: One of the biggest threats to our environment today is the excessive waste. Lean manufacturing is an approach that has been most adaptable to its environment. It can be applied without negative effects on the working environment.
- *Competitiveness:* When properly applied, lean can create huge improvements in efficiency, cycle time, and productivity. Today, many organizations adopt lean manufacturing strategy that would help them compete in the competitive globalization market.
- *Customer Satisfaction:* Lean manufacturing proposes a customer-value focus to achieve the high efficiency. Lean involves only producing a product when the customer wants it. So a company needs to know their customers and identify their greatest need. Since lean implies adding value to processes and services, value means any useful activities for which the customer is willing to pay.
- *Workplace Efficiency:* Lean manufacturing improves workplace efficiency. Inventory and employee's skills are used more efficiently. It is recognized as one of the most efficient and effective global operation strategies.

Figure 11.5 depicts some of these benefits.

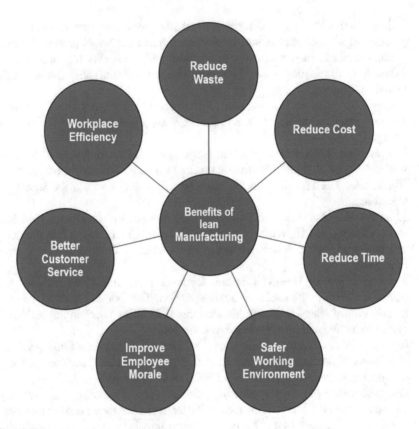

Fig. 11.5 Benefits of lean manufacturing

11.9   Challenges

Only a few companies can assess the impact of lean manufacturing at an early stage in order to determine its viability and profitability. Lean manufacturing can suffer if management decides to implement it without first consulting with its employees. There is lack of standard LM implementation process/framework. LM practices are regarded as advanced manufacturing practices and more than 50% of lean manufacturing implementation efforts fail. The major issues that make lean implementations a challenging task include time constraint, constraints of product quality, and complexity of the manufacturing. A rush to become lean without adequate understanding has resulted in several misapplications of lean manufacturing tools [32, 33]. Other challenges include the following:

- *Lack of Commitment*: Lean is perceived negatively, and organizations fail to dedicate time, manpower, and other resources toward lean. Implementing lean manufacturing causes a profound cultural change that will alter the role of every person in entire organization. There is a lack of commitment from senior and middle management and poor understanding of lean manufacturing concepts.

- *Lack of Change Management:* The success of lean manufacturing requires consistent and conscious efforts from the entire organization. Implementing lean requires everyone to be on board. Lack of support from senior management will cause lean implementations to fail.
- *Lack of Teaching Materials:* Successful implementation of lean principles requires that managers and employees be well-informed and educated in the application of lean tools. However, there is a shortage of lean teaching materials available to both corporate and academic educators.
- *Bad Name:* Lean has gotten a bad name since its inception in the 1990s. It has become a euphemism: "We cut jobs 10% to lean our organization." That view is antithetical to lean thinking, which relies on creating trust for the long-term relationships needed to transform an organization. "There is a danger that companies associate lean with quick wins and lose the long-term benefits."
- *Standardization:* Standardization creates the conditions necessary for controlled scientific inquiry. Demanding standardization in all aspects of production makes it easy for workers and managers to immediately identify problems.

11.10   Global Adoption of Lean Manufacturing

Lean manufacturing basically aims at eliminating any waste in all production phases in order to ensure their survival, obtain high productivity, excellent product quality, and reduced manufacturing costs. Many companies around the world are drawn to the idea of lean manufacturing. Lean manufacturing principles have been recognized as one of the most efficient global operation strategies. They have emerged as a powerful approach that has been used by companies in developing countries to improve their operations. Developing countries such as India, Kuwait, Malaysia, Turkey, Brazil, Thailand, Zimbabwe, South Africa, and Indonesia have adopted the philosophy to reduce manufacturing costs. We now consider how lean manufacturing is implemented in different nations.

- *Japan:* Lean manufacturing is American name to Toyota Production System. It is often regarded as the Toyota Way since it was developed by the company. In the Toyota Way, improving workflow is the main goal, and in so doing waste is also eliminated naturally. To achieve the goal, Toyota applies a mentoring methodology called Senpai and Kohai, which translates to senior and junior. The Toyota Production System (or The Toyota Way) led by Taiichi Ohno and Shigeo Shingo is grounded by the following six principles: eliminating waste, streamlining processes, increasing efficiency, improving productivity, respecting people, and pleasing the customer. The Toyota Production System is illustrated in Fig. 11.6 [25]. It operates on the following 14 principles [34–36]:

1. Basing management decisions on long-term philosophy.
2. Revealing problems needing to be fixed by creating a continuous process flow.
3. Avoiding overproduction by using a "Pull" system.

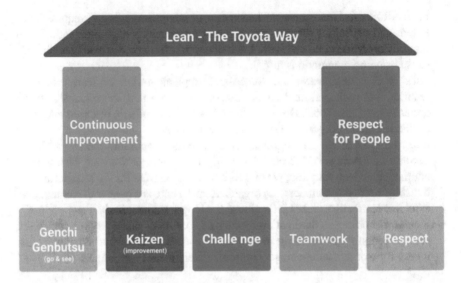

Fig. 11.6 Toyota Production System House [25]

4. Leveling out workloads.
5. Producing quality products the first time by implementing a continuous improvement culture.
6. Standardizing tasks.
7. Using visual controls.
8. Using reliable technology.
9. Encouraging the growth of leaders who will pass on knowledge.
10. Developing people that will follow the company's philosophy.
11. Respecting partners by challenging them and helping them improve.
12. Using "Gemba".
13. Making consensus decisions.
14. Becoming a learning organization.

These principles have been widely regarded as the system that made Toyota as successful as it is today [34].

- *United States:* An increasing number of US companies are outsourcing manufacturing operations to foreign countries with cheaper labor markets. The issues of keeping manufacturing jobs in America and increasing the competitiveness of the US manufacturing industry have become critical to the long-term sustainable prosperity of the US economy. Lean manufacturing has proven to be an effective strategy to increase productivity and cost competitiveness [37]. Henry Ford is commonly referenced as the true starting point of lean, and the Ford Production System paved the way for most modern lean manufacturing and is one of the best examples of lean manufacturing principles.

- *United Kingdom:* UK companies have been enthusiastic in employing lean principles. The lean approach to business is focused on delivering customer value using the least resource. Many production processes have applied the lean strategy with waste removed to bring efficiency gains [38]. For example, Oxford Engineering introduced lean principles throughout its operation, including its administrative functions. As a result, the company delivered a 16% increase in productivity.
- *India:* The traditional manufacturing industry has changed rapidly in developing nations like China or India in search of cheap labor and more profit. There are so many companies now producing goods in China and India that simply relying on their low-cost labor. India is entering the global market and emerging as the new destination for global manufacturing. India is the place to be for design, development, and manufacturing of innovative products and major companies from Europe, the United States, and Japan [29].
- *Zimbabwe:* The manufacturing sector in Zimbabwe has been struggling in their operations due to inadequate funding to improve on its machinery and technology. This has forced manufacturing companies in Zimbabwe to implement lean manufacturing in order to eliminate waste and improve the product quality. The main strength of lean manufacturing implementation in Zimbabwe is that the employees got motivated about the program which made them to be dedicated and hardworking. Its main weakness is that the implementation created fear among the workers for job losses through retrenchment [39].

11.11 Conclusion

Simply stated, lean manufacturing or "going lean" is a philosophy based on a strategic approach of continuous improvement through the elimination of waste and inefficiencies. It is a systematic approach to identifying the value-adding activities and then eliminate the non-value-adding activities. It involves using several techniques to remove or reduce waste from production. It is the latest, most well-known, and most successful methodology being used by companies to turn their business around. It deserves the full attention of both employers and employees, because its objectives are to improve productivity, reduce waste, increase competitiveness, and optimize available resources. Companies, academic institutions, and governments are pushing their own version of lean as the strategy to manufacturing, educational, governmental salvation. Lean has now become a business system or culture.

There is a shortage of lean teaching materials available to both corporate and academic trainers. Some institutions have started integrating lean manufacturing into their traditional manufacturing engineering programs, and they are preparing students to better understand and practice lean manufacturing [33]. More information about lean manufacturing can be found in the books in [36, 40–51].

References

1. M.N.O. Sadiku, O.D. Olaleye, S.M. Musa, Lean manufacturing: A primer. Int. J. Trend Res. Dev. **7, no. 1**, 173–176 (2020, January–February)
2. J. Bhamu, K.S. Sangwan, Lean manufacturing: Literature review and research issues. Int. J. Oper. Prod. Manag. **34**(7), 876–940 (2014)
3. P. Rewers, J. Trojanowska, P. Chabowski, Tools and methods of lean manufacturing—A literature review, in *Proceedings of 7th International Technical Conference Technological Forum*, Czech Republic, 2016, pp. 135–139
4. 7 Types of waste in lean manufacturing, https://www.planettogether.com/blog/seven-types-of-waste-in-lean-manufacturing
5. Steps to reduce waste with lean manufacturing, https://www.mimeo.com/blog/reduce-waste-lean-manufacturing/
6. S. Gupta, S.K. Jain, A literature review of lean manufacturing. Int, J. Manag. Sci. Eng. Manag. **8**(4), 241–249 (2013)
7. Lean manufacturing, *Wikipedia*, the free encyclopedia, https://en.wikipedia.org/wiki/Lean_manufacturing
8. J.P. Womack, D.T. Jones, *Lean Thinking: Banish Waste and Create Wealth in your Corporation*, 2nd edn. (Free Press, 2003)
9. B.M. Kariuki, D.K. Mburu, Role of lean manufacturing on organization competitiveness. Ind. Eng. Lett. **3**(10), 81–92 (2013)
10. A lean and green approach to reduce waste, improve quality, https://www.foodengineering-mag.com/articles/98301-a-lean-and-green-approach-to-reduce-waste-improve-quality
11. Lean manufacturing 101, https://blog.unex.com/lean-manufacturing-101
12. G. Immerman, Lean manufacturing: A complete guide to lean for manufacturers (2021, February), https://www.machinemetrics.com/blog/lean-manufacturing
13. M.N.O. Sadiku, K.G. Eze, S.M. Musa, Six Sigma: An introduction. Int. J. Trend Res. Dev. **7**(2), 191–194 (2020, March–April)
14. Lean Six Sigma, *Wikipedia*, the free encyclopedia, https://en.wikipedia.org/wiki/Lean_Six_Sigma
15. S.K. Singh et al., Role & importance of lean manufacturing in manufacturing industry. Int. J. Eng. Sci. **3**(6), 1–14 (2014)
16. M.N.O. Sadiku, S.M. Musa, O.M. Musa, The essence of industry 4.0. Invent. J. Res. Technol. Eng. Manag. **2**(9), 64–67 (2018, September)
17. J.A.H. Kareem, P.S.M. Al Askari, F.H. Muhammad, Critical issues in lean manufacturing programs: A case study in Kurdish iron & steel factories. Cogent Eng. **4**(1) (2017)
18. V. Gupta, P. Mota, Lean manufacturing: A review. Int. J. Sci. Technol. Manag. **3**(2), 176–180 (2015, July)
19. Lean manufacturing, https://www.sciencedirect.com/topics/engineering/lean-manufacturing
20. P. Dhiravidamani et al., Implementation of lean manufacturing and lean audit system in an auto parts manufacturing industry – An industrial case study. Int. J. Comput. Integr. Manuf. **31**(6), 579–594 (2018)
21. T.N. Issa, Lean manufacturing implementation in fused plastic bags industry. Int. J. Indust. Manuf. Eng. **12**(10), 1151–1158 (2018)
22. T. Shimizu, Applications of lean manufacturing in the machine tool industry, Master's Thesis, Northern Kentucky University, 2004, July
23. M. Engum, Implementing lean manufacturing into newspaper production operations, Master's Thesis, Rochester Institute of Technology, 2009, May
24. M.S. Obeidat, R. Al-Aomar, Z.J. Pei, Lean manufacturing implementation in the sewing industry. J. Enterp. Transform. **4**(2), 151–171 (2014)
25. What is lean management? Definition & benefits, https://kanbanize.com/lean-management/what-is-lean-management

26. C.H. Li, H.K. Lau, Application of lean manufacturing in product safety, in *Proceedings of IEEE Symposium on Product Compliance Engineering,* 2018, December
27. P. Hasle, A.P. Nielsen, K. Edwards, Application of lean manufacturing in hospitals—The need to consider maturity, complexity, and the value concept. Hum. Factors Ergon. Manuf. Serv. Indust. **26**(4), 430–442 (2016)
28. R. Shah, P.T. Ward, Lean manufacturing: Context, practice bundles, and performance. J. Oper. Manag. **21**, 129–149 (2003)
29. M. Ghosh, Lean manufacturing performance in Indian manufacturing plants. J. Manuf. Technol. Manag. **24**(1), 113–122 (2013)
30. Why companies use lean manufacturing, https://gesrepair.com/why-companies-use-lean-manufacturing/
31. P. Landau, What is lean manufacturing? (2019, May), https://www.projectmanager.com/blog/what-is-lean-manufacturing
32. S.J. Pavnaskar, J.K. Gershenson, A.B. Jambekar, Classification scheme for lean manufacturing tools. Int. J. Prod. Res. **41**(13), 3075–3090 (2003)
33. P.W. Shannon, K.R. Krumwiede, J.N. Street, Using simulation to explore lean manufacturing implementation strategies. J. Manag. Educ. **34**(2), 280–302 (2010)
34. The complete lean manufacturing guide, https://tulip.co/ebooks/lean-manufacturing/
35. Lean manufacturing, https://www.creativesafetysupply.com/articles/lean-manufacturing/
36. J. Liker, *The Toyota Way: 14 Management Principles from the World's Greatest Manufacturer* (McGraw-Hill Education, 2004)
37. N. Fang, R. Cook, K. Hauser, Work in progress: An innovative interdisciplinary lean manufacturing course, in *Proceedings of 36th ASEE/IEEE Frontiers in Education Conference,* October 28–31, 2006, San Diego, CA, 2006, October
38. G. Parry, Lean manufacturing. IEE Manuf. Eng., 44–47 (2005, October/November)
39. C. Maware, O. Adetunji, Lean manufacturing implementation in Zimbabwean industries: Impact on operational performance. Int. J. Eng. Bus. Manag. **11**, 1–12 (2019)
40. K.D. Zylstra, *Lean Distribution: Applying Lean Manufacturing to Distribution, Logistics, and Supply Chain* (Wiley, Hoboken, NJ, 2006)
41. J. Page, *Implementing Lean Manufacturing Techniques: Making your System Lean and Living with IT* (Hanser Gardner Publications, 2004)
42. L.E. Fast, *The 12 Principles of Manufacturing Excellence: A Lean Leader's Guide to Achieving and Sustaining Excellence,* 2nd edn. (CRC Press, Boca Raton, FL, 2016)
43. W.M. Feld, *Lean Manufacturing: Tools, Techniques, and How to Use Them* (CRC Press, Boca Raton, FL, 2000)
44. K.D. Zylstra, D. Kirk, *Lean Distribution: Applying Lean Manufacturing to Distribution, Logistics, and Supply Chain* (Wiley, Hoboken, NJ, 2006)
45. J.P. Davim (ed.), *Progress in Lean Manufacturing* (Springer, 2018)
46. D.P. Hobbs, *Lean Manufacturing Implementation: Complete Execution Manual for Any Size Manufacturer* (J. Ross Publishing, 2004)
47. L. Wilson, *How to Implement Lean Manufacturing* (McGraw-Hill, New York, 2010)
48. G. Conner, *Lean Manufacturing for the Small Shop,* 2nd edn. (Society of Manufacturing Engineers, 2008, October)
49. M. Dudbridge, *Handbook of Lean Manufacturing in the Food Industry* (Blackwell Publishing, 2011)
50. F.J.G. Silva, L.C.P. Ferreira (eds.), *Implementation, Opportunities and Challenges* (Nova Science Publishers, 2019)
51. Y. Koren, R. Hill, *The Global Manufacturing Revolution: Product-Process-Business Integration and Reconfigurable Systems* (Wiley, Hoboken, NJ, 2010)

Chapter 12
Distributed Manufacturing

> *A potato can grow quite easily on a very small plot of land.*
> *With molecular manufacturing, we'll be able to have distributed*
> *manufacturing, which will permit manufacturing at the site*
> *using technologies that are low-cost and easily available.*
> —Ralph Merkle

12.1 Introduction

Distributed manufacturing (DM) refers to the decentralized fabrication of parts in smaller factories or homes that are local to end users. It makes it possible for more innovators to reach the market with their products, while manufacturing is the process by which materials are transformed into usable products.

Distributed manufacturing contributes to the inputs needed by other industries by providing products ranging from heavy-duty machinery to home electronics. The traditional manufacturing industry has focused mainly on mass production models to raise productivity. It has been characterized by the progressive centralization of operations that focus on benefits from economies of scale. It is responsible of a large share of the global environmental impacts as it releases significant amounts of emissions and waste [1]. It works well for large organizations, but it is not as friendly to small businesses with few products on the market. It requires an expensive and time-consuming effort to develop and maintain. New architectures are needed for a new manufacturing paradigm to meet these challenges. These architectures must be agile, reliable, configurable, and adaptable [2].

Leveraging the advantages of Industry 4.0 and its related technologies such as cloud computing, the Internet of Things, and 3D printing has made distributed manufacturing fascinating. This manufacturing-as-a-service model is achieved by making small orders and prototyping affordable and building their production facilities near the places of consumption; consequently, today's new manufacturing paradigm is more distributed than ever [3].

© The Author(s), under exclusive license to Springer Nature Switzerland AG 2023
M. N. O. Sadiku et al., *Emerging Technologies in Manufacturing*,
https://doi.org/10.1007/978-3-031-23156-8_12

This chapter provides a further introduction to distributed manufacturing. It begins by explaining the concept of distributed manufacturing. It describes distributed manufacturing systems and the features of distributed manufacturing. It covers some applications of DM. It highlights the benefits and challenges of DM. It addresses the global adoption of DM. The last section concludes with comments.

12.2   Concept of Distributed Manufacturing

Distributed manufacturing (DM) is all about creating products as close as possible to the customers who will use them. In traditional manufacturing, raw materials are often assembled and fabricated in large, centralized factories, and the finished products are then distributed to the customer. A manufacturer will own dedicated factories to create a convenient supply chain for their clients. Distributed manufacturing turns this approach around. In distributed manufacturing (DM), the raw materials and fabrication techniques are decentralized so that the final product is fabricated very close to the final customer. In contrast to mass manufacturing, distributed manufacturing entails fabricating products in many locations around the world, close to where they are needed. Distributed manufacturing is a form of decentralized manufacturing which disrupts the scale economics of traditional manufacturing.

In today's globalized world economy and the rapid development of information and communication technologies (ICT, i.e., the infrastructure and components that enable modern computing), the manufacturing resources of an enterprise are usually distributed geographically. Distributed manufacturing involves a complete rethinking of what it means to be a manufacturer. Raw materials remain decentralized, while final assembly occurs on a much more individualized basis, close to the customer. In other words, components may be manufactured in different physical locations and then are brought together for assembly of the final product, near point of use locations. It means making parts all over the world and using supply chain management concepts to bring it all together for final production near the point of sale or use jurisdictions.

12.3   Distributed Manufacturing Systems

Distributed manufacturing systems (DMS) represent an ideal approach to meeting challenges regarding the individualization of products, predicated on customer proximity, or more sustainable production. They may be regarded as a class of manufacturing systems, focused on internal manufacturing control and characterized by such common properties as autonomy, flexibility, adaptability, agility, and decentralization.

As shown in Fig. 12.1, the concept of DMS evolved from a decentralized form of production control to a form of manufacturing system involving several enterprises

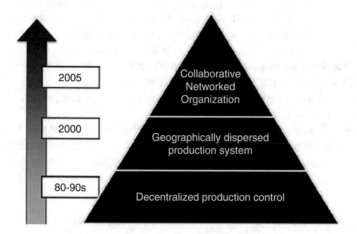

Fig. 12.1 Evolution of DMS concept [4]

[4]. Initially, DMS models were a form of decentralized production systems. Then, DMS were characterized by geographical dispersion of the production systems. Later, DMS evolved to the status of manufacturing activities among several closely proximate enterprises, aided by information communication technologies (ICTs) linked by geographically distributed control systems.

12.4 Features of Distributed Manufacturing

Distributed manufacturing is geographically distributing manufacturing facilities so that companies can get competitive advantages regarding costs, time, and quality. Distributed manufacturing has the following features [5].

- *Localization:* Manufacturing companies have realized that market and customer proximity have become an important factor for success on the market. Distributed manufacturing enables the local production of goods. Producing locally, close to the consumer, eliminates transportation costs, taxes, and tariffs. It also reduces emissions and negative environmental impacts. Adopting the distributed manufacturing system makes moral sense to the local community.
- *Customization:* A major characteristic of DM is the ability to produce customized products. DM can produce products that are individually tailored to very small markets. Customization and enhanced customer satisfaction is the answer to an increasing individualization of customer request.
- *Personalization:* In today's competitive global market, customers demand personalized products. Distributed manufacturing has advantages over traditional manufacturing in that it can produce personalized product offerings which enhances customer loyalty and invites repeat business opportunities. Consumers can submit their own personalize products and are more involved in the design and fabrication of products.

- *Product Longevity:* Distributed production can increase product longevity through proximate repair centers and enhanced repair speeds. With 3D printing, spare parts can be easily made with lower customization costs.
- *Easy Prototyping:* With distributed manufacturing, device and PCB prototyping should be quick, reasonably priced, and painless.
- *Low Inventory:* Distributed manufacturing aims at keeping inventories low in favor of just-in-time production.
- *Efficient Use of Resources:* Distributed manufacturing is expected to enable a more efficient use of resources, with less wasted capacity in centralized factories. It should reduce the overall negative environmental impact of manufacturing.

12.5   Applications of Distributed Manufacturing

Distributed manufacturing is essentially about creating products as close as possible to the customers. The most obvious changes to come from distributed manufacturing involve production processes. Information flow between the shop floors that are combined in a production network in a geographically distributed manufacturing shop is shown in Fig. 12.1 [6].

- *On-Shore Manufacturing:* Offshore manufacturing presents serious challenges, such as language barriers, time zones, enhanced opportunities for compromised quality and reliability, and many layers of sales and logistics. In the past, the only way to get electronics product manufacturing accomplished at a reasonable price is to do it offshore. With the rise of distributed manufacturing, there are now many options in the United States for cost-effective, reliable, and fast production and delivery realizations. On-shore manufacturing saves organizations the added complexity and risk associated with overseas operations.
- *Furniture Industry:* A company like AtFAB, a US maker of wood furniture, uses distributed manufacturing to meet customer demands. It replaces the aggregation of wood materials in a single factory with the distribution of computerized numerical control files to localized manufacturing sites. Parts are then assembled by the consumer or by local fabrication workshops that can turn them into finished products.
- *Sustainable Construction:* In recent years, there has been a shift toward more sustainable approaches for construction. Lean construction is regarded as one of the foundational viewpoints for the consideration of environmental factors within construction. Minimization of waste is of prime consideration in lean construction. In terms of construction, lean principles focus on the industry-wide reduction of waste. Sustainable construction seeks to reduce the environmental footprints of building sites. Sustainable modular construction can be viewed in terms of the distributed manufacturing paradigm [7].
- *Desktop Factories:* These are the new form of distributed manufacturing system, emerging through the recent progress in additive manufacturing (or 3D-printing).

They play significant roles in supporting the on-demand manufacturing of customized products. Such small-scale production systems show an ideal approach for DIY (do-it-yourself) manufacturing [8].

- *Product-Service System* (PSS): This concept is regarded as a promising type of business models to improve production and consumption toward social, economic, and environmental sustainability. It is an integrated offering of products and services that fulfill specific customer demand. Distributed manufacturing, a network of small-scale production units equipped with advanced manufacturing technologies that are proximate to the point of sale and use, can be applied to PSS to address its implementation barriers such as organizational, financial, technical, social, cultural, and regulatory barriers [9].

12.6 Benefits

If managed properly, distributed manufacturing can offer significant benefits, including
reduced manufacturing costs, short time-to-market, increased supply-chain flexibility, increased product safety, quality, reliability, and efficient use of resources. It allows technology designers and implementers to choose from a variety of machines, materials, and quantities with unparalleled flexibility, quality, cost-effectiveness, reliability, production on demand, and speed. Some manufacturers believe they have a lot to gain by distributing their products directly to the point of consumption and bypassing their intermediaries. Manufacturers can support multiple smaller economies by distributing their factories. Other benefits include [10]:

- *Efficient Use of Resources:* Distributed manufacturing will enable a more efficient use of resources, with less wasted capacity in centralized factories. It should reduce the overall negative environmental impact of manufacturing. It is capable of delivering superior results in five (5) key ways: speed, quality, reliability, safety, and cost-effectiveness.
- *Localization:* Implementing DM-oriented techniques will achieve digitalization, personalization, and localization. Distributed manufacturing enables the local production of goods. Producing locally, close to the consumer, eliminates transportation costs, taxes, and tariffs. By manufacturing products closer to their end destination, we reduce time to the market, logistics costs, environmental costs, and associated impacts.
- *Friendly to Small Business*: Small businesses with few products on the market do not always have the demand required to sustain or support traditional manufacturing. DM is friendly to small businesses or newly developed products by making both small orders and rapid-prototyping affordable. Employing DM can lead to faster growth.
- *Reduced Costs*: Many of the capabilities of distributed manufacturing led to reduced costs. This is a key motivation for small businesses that are introducing products to market on a tight budget. DM also requires significantly less up-front

capital than traditional manufacturing. Logistical costs and environmental impacts are minimized when products are made geographically closer to their end user destinations.

- *Faster Time-to-Market:* By deploying the whole manufacturing process with user-friendly online tools and rapid prototyping, distributed manufacturing helps an enterprise get products into the hands of buyers more quickly and reduce transit time with fewer logistical challenges.
- *Superior Quality:* High-volume manufacturers often lose quality, while distributed manufacturing enables small shops to produce consistent and high-quality work and resulting products.
- *Customer Relationships:* Distributed manufacturing requires a different approach to maintaining relationships with customers. You are not flooding the market with large quantities of products but are producing demand-tailored goods that are closer to the market.

12.7   Challenges

Distributed manufacturing is not without its challenges. The process of adopting to such fundamental shifts in manufacturing is daunting. DM may be more difficult to regulate and control remotely. Not everything can be made via distributed manufacturing. A major challenge of DM is the ability to upscale while retaining the value that the model aims to create [11, 12]. Other challenges include the following (Figs. 12.2 and 12.3):

- *Workforce:* Workers will no longer be centralized in large facilities. Instead, they will operate at smaller, distributed sites. They will rely increasingly on electronic files and mobile apps. Also, they will require entirely new skill sets. That means recruiting, training, and retaining a very different workforce.
- *Complexity*: Due to the complexity of products created in distributed manufacturing, collaboration and coordination are required among several facilities typically operating in different time zones. There is also the complexity of producing identical parts in separate facilities around the globe and with different standards and regulatory requirements.
- *Security:* This must be paramount to protect the intellectual property rights of the organization in all distributed transactions.

12.8   Global Adoption of Distributed Manufacturing

The primary benefit for adopting distributed manufacturing is the ability to create value at geographically dispersed locations. DM allows organizations in cities and small towns affected by globalization and outsourced manufacturing to produce

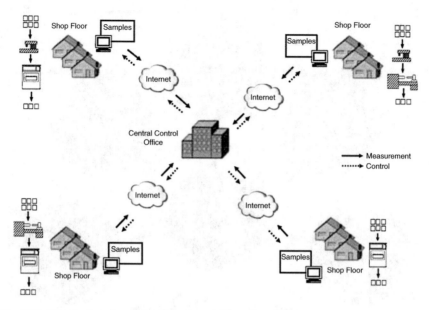

Fig. 12.2 The distributed manufacturing shops [6]

parts with less capital and overhead than ever before. Startups and established organizations around the world are now exploring the opportunities that distributed manufacturing presents. We now consider the adoption of distributed manufacturing in some nations [13].

- *United States:* For continued growth, US manufacturers must continually disrupt themselves and keep a mindset of innovation. Most factories in the United States operate at far less than capacity. Nike, for instance, is already localizing its manufacture to optimize resource use by reducing the amount of land needed for large factories and saving on fuel and logistics for transportation. Nike could be a pure software company in the next ten years, thanks to 3D printing production.
- *China:* China has made tremendous strides since opening its economy 30 years ago. China has been undergoing a transformation, moving from being the world's manufacturing base to becoming a more integrated economy with a wider variety of value-added capabilities. The shift from being a manufacturing and export-oriented economy to a more integrated one is well recognized by policymakers in China. These changes are occurring because China's cost-based advantages are being eroded through economic growth. The consequence has slowed down economic growth in China. Even with this slowdown, many analysts and policymakers still see China as a strong market for domestic and foreign investment. The drop-off has been attributed by some to a lack of confidence in the short-term in China's economy. The base population of 1.3 billion people and the huge market continue to make China the most promising market among all the world's emerging economies. At the national level, according to President Hu

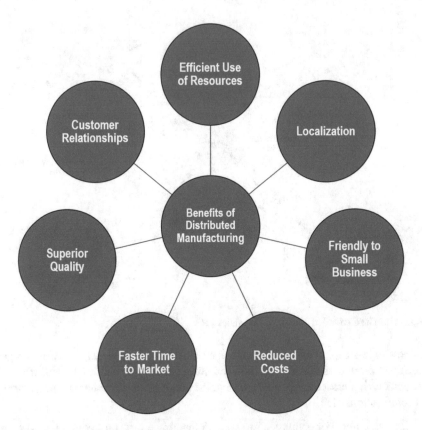

Fig. 12.3 Some benefits of distributed manufacturing

Jintao, the government has been paying more attention to environmental issues for the purpose of sustainable development.

- *Singapore*: Asia is the fastest growing market region with many emerging economies currently booming and growing at a very healthy state. The adoption of 3D-printed products being used is a very encouraging indicator. In Singapore, companies have been co-innovating with customers to deliver new solutions. This collaboration with 3D Printing Studios is part of the goal to support manufacturers and adopt scalable business process. In this partnership, 3D Printing Studios seeks to serve B2B customer bases across 12 countries, including expansion in Singapore and Australia and beyond to New Zealand, Malaysia, Indonesia, Thailand, Philippines, Vietnam, India, Sri Lanka, Japan, and South Korea.
- *Spain:* The city of Barcelona has launched a Fab City project that aims to restore production centers to the heart of the city. The plan is to open fab labs in every district of the city, and eventually every neighborhood, to enable local production of almost any kind of goods. The fab labs are also planned to be community problem-solving centers that assist in local energy and food production.

12.9   Conclusion

Manufacturing is becoming more complex and distributed than ever. Distributed manufacturing represents a radically different model of consumer goods production, purchase, and use. Many companies have recently changed their manufacturing system from traditional single factory to multi-factory to increase their international competitiveness. Using this approach, factories processing various machines and tools at different geographical locations are often combined [14, 15].

Distributed manufacturing is regarded as one of the leading emerging technologies for the future. It is the future of how products are fabricated and delivered to customers who use them. Interest in distributed manufacturing is progressively growing as it incorporates new technological advancements such as the Internet of Things (IoT), cloud computing, cloud manufacturing, semantic web, and virtualization [16]. The idea of distributed manufacturing by geographically distributed production will continue to play an important role. More information about distributed manufacturing can be found in the books in [17–20].

References

1. F. Cerdas et al., Life cycle assessment of 3D printed products in a distributed manufacturing system. *Journal of Industrial Ecology* **21**(S1) (2017)
2. F.P. Maturana, D.H. Norrie, Multi-agent mediator architecture for distributed manufacturing. *Journal of Intelligent Manufacturing* **7**, 257–270 (1996)
3. E. Rauch et al., Sustainability in manufacturing through distributed manufacturing systems (DMS). *Procedia CIRP* **29**, 544–549 (2015)
4. M. Seregni, C. Zanetti, M. Taisch, Development of distributed manufacturing systems (DMS) concept, http://www.summerschool-aidi.it/edition-2015/images/Naples2015/proceed/22_seregni.pdf
5. M. Kumar et al., Developing distributed manufacturing strategies from the perspective of a product-process matrix. *International Journal of Production Economics* **219**(January), 1–17 (2020)
6. I. Mahdavi et al., Designing an E-based real time quality control information system for distributed manufacturing shops, in *Proceedings of the 9th International Conference on Enterprise Information Systems,* Madeira, Portugal, 2007, June
7. C. Turner, J. Oyekan, L.K. Stergioulas, Distributed manufacturing: A new digital framework for sustainable modular construction. *Sustainability* **13** (2021)
8. E. Rauch et al., Collaborative cloud manufacturing: Design of business model innovations enabled by cyberphysical systems in distributed manufacturing systems. *Journal of Engineering* (2016)
9. P. Aine, A design tool to apply distributed manufacturing principles to sustainable product-service system development, https://pdfs.semanticscholar.org/aa2c/0900a549bdb689ea12c09f281d460ee05a08.pdf?_ga=2.168324258.220029878.1625074349-483007315.1622922079
10. Distributed manufacturing aims to replace much of the material supply chain with digital information, https://macrofab.com/distributed-manufacturing/
11. J.S. Srai et al., Distributed manufacturing: Scope, challenges and opportunities. *International Journal of Production Research* **54**(23), 6917–6935 (2016)

12. R. Hayfords, 5 keys to success in a distributed manufacturing model (2020, August), https://www.3yourmind.com/news/5-keys-to-success-in-a-distributed-manufacturing-model
13. B. Jackson, Manufacturing (2017, September), https://3dprintingindustry.com/news/3d-printing-studios-partners-sap-distributed-manufacturing-121966/
14. W. Zhang, M. Gen, Process planning and scheduling in distributed manufacturing system using multiobjective genetic algorithm. *IEEJ Transactions on Electrical and Electronic Engineering* **5**, 62–72 (2010)
15. M.N.O. Sadiku, P.O. Adebo, A. Ajayi-Majebi, S.M. Musa, Distributed manufacturing. *International Journal of Trend in Research and Development* **7**(6), 150–151 (2020, November–December)
16. G. Adamson, L. Wang, P. Moore, Feature-based control and information framework for adaptive and distributed manufacturing in cyber physical systems. *Journal of Manufacturing Systems* **43**, 305–315 (2017)
17. L. Wang, W. Shen, *Process Planning and Scheduling for Distributed Manufacturing* (Springer, London, 2010)
18. ESPRIT Consortium CCE-CNMA, *CCE: An Integration Platform for Distributed Manufacturing Applications: A Survey of Advanced Computing Technologies* (Springer, Berlin, 1995)
19. H. Kuhnle, G. Bitsch, *Foundations & Principles of Distributed Manufacturing* (Springer, Cham, 2015)
20. H. Kuhnle (ed.), *Distributed Manufacturing: Paradigm, Concepts, Solutions and Examples* (Springer, London, 2009)

Chapter 13
Cloud Manufacturing

The cloud services companies of all sizes...The cloud is for everyone. The cloud is a democracy.

–Marc Benioff.

13.1 Introduction

The manufacturing industry is crucial to the economy of any nation. Manufacturing is a vital sector in our society. It has always been a pillar industry of developed economies. It has been a strong impetus to the process of industrialization and modernization. Today's manufacturing sector includes a wide range of industries, from food to automobile and aerospace and from computer to electronic product manufacturing [1].

Modern manufacturing has changed significantly due to intense global competition and remarkable advances in information technology. To survive in an increasing competitive pressure, globalization, and rapid technology development, modern manufacturing requires a flexible and dynamic provisioning and management, using manufacturing resources available on demand over computer networks. Technology has always been the main driver of the transformation of manufacturing. Information technology is transforming the manufacturing industry by digitizing virtually every facet of modern manufacturing processes. Today, manufacturing is shifting from production-oriented manufacturing to service-oriented manufacturing [2]. Modern manufacturing system and processes are becoming increasingly complex, requiring users to use advanced tools such as cloud computing.

Cloud manufacturing (CM) (also known as Manufacturing-as-a-Service (MaaS) is newly emerging as a new promising manufacturing paradigm developed from existing advanced manufacturing models. It is a recent approach of adopting well-known basic concepts from cloud computing to manufacturing processes and deliver-shared, ubiquitous, on-demand manufacturing services. It refers to a customer-centric approach for enabling ubiquitous, convenient, on-demand network access to a shared pool of manufacturing resources and capabilities that can be rapidly provisioned and released with minimal management effort. CM transforms manufacturing resources and abilities into a cheap and on-demand virtualized service pool (manufacturing cloud) for the whole life cycle of a product (e.g., design, simulation, production, test, verification, redesign, reuse, maintenance). In a CM system, manufacturing resources can be connected into the Internet and managed using the Internet of Things (IoT) technologies such RFID, wired and wireless sensor network, and embedded devices [3].

This chapter provides an introduction to cloud manufacturing. It begins by providing an overview of cloud computing. It explains cloud manufacturing. It describes some of the applications of cloud manufacturing. It highlights the benefits and challenges of CM. It covers global cloud manufacturing and the future of cloud manufacturing. The last section concludes with comments.

13.2 Overview of Cloud Computing

"Cloud" is a metaphor for the Internet. Cloud computing (CC) represents the underlying platform technology enabling cloud manufacturing. It is a newly emerging service-oriented computing technology. It is the provision of scalable computing resources as a service over the Internet. It is a smart network manufacturing model that enables service-oriented personalized manufacturing. It allows manufacturers to use many forms of new production systems such as 3D printing, high-performance computing (HPC), industrial Internet of Things (IIoT), and industrial robots. It is transforming virtually every facet of modern manufacturing [4].

Cloud computing has been viewed as an innovation in computing whose key feature is the virtualization of computing resources and services. All kinds of manufacturing resources (including machines, processing centers, and computing equipment) and capacities will be virtualized, unified, and centralized for management and operation, to achieve an intelligent multi-win-win situation. This extensive networked environment of resources will provide high added value, low-cost products, and global manufacturing services [5].

Cloud computing provides shared computer resources and data to users through the Internet. It enables ubiquitous, convenient, on-demand network access to share pool of computing resources such as networks, servers, storages, and services. It aims at making computing a utility such as water, gas, electricity, and telephone services. Computation, which used to be confined to one location, is now centralized in vast shared facilities [6].

The key characteristic of cloud computing is the virtualization of computing resources and services. In cloud computing, everything is treated as a service (i.e., XaaS). There are three major services: software as a service (SaaS), platform as a service (PaaS), and infrastructure as a service (IaaS). These services are illustrated in Fig. 13.1 and explained as follows [7, 8].

- *SaaS:* This is a software delivery model in which software and associated data are hosted on the cloud. In this model, cloud service providers offer on-demand access to computing resources such as virtual machines and cloud storage. This

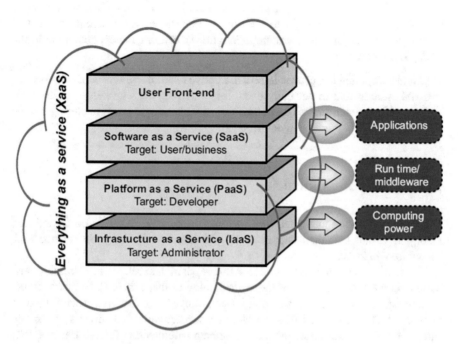

Fig. 13.1 In cloud computing, everything is a service [7]

is also known as "software on demand." For example, a cloud-based enterprise resource planning (ERP) for manufacturing is a type of SaaS used in the industry.

- *PaaS:* The platform is provided as a service. This allows the end user to create a software solution using tools or libraries from the platform service provider. In this model, cloud service providers deliver computing platforms such as programming and execution.

- *IaaS:* This provides infrastructure hardware resources (such as computers, storage, servers, etc.) as a service and allows users to customize their own IT infrastructure dynamically.

Just like cloud computing, CM services can be categorized into four major deployment models (public, private, community, and hybrid clouds) [9, 10]:

- *Private cloud* refers to a centralized management effort in which manufacturing services are shared within one company or its subsidiaries. A private cloud is often used exclusively by one organization, possibly with multiple business units.

- *Community cloud* is a collaborative effort in which manufacturing services are shared between several organizations. Services are provided to multiple organizations from a certain community with similar business goals.

- *Public cloud* realizes the key concept of sharing services with the general public. Public clouds are commonly implemented through data centers operated by providers such as Amazon, Google, IBM, and Microsoft.

- *Hybrid cloud* essentially spans multiple configurations and is composed of two or more clouds (private, community, or public), offering the benefits of multiple deployment modes.

Cloud computing has proven to be a disruptive technology. It is becoming pervasive and omnipresent in our daily lives. It is changing the way industries do their businesses. It is emerging as one of the major enablers for the manufacturing industry. It has inspired cloud manufacturing [7].

13.3   Cloud Manufacturing

The concept of cloud manufacturing was first proposed in 2009 by Bo Li and Lin Zhang in China and then supported by the Chinese Ministry of Science and Technology in 2010.

The term "cloud" means that your software, data, and related infrastructure are hosted via the Internet. State-of-the-art technologies that enable CM include cloud computing, pay-as-you-go/use utility computing, virtualization, social media, Internet of Things (IoT), RFID, wireless sensor network, embedded system, big data, wearable technologies, and service-oriented architecture (SOA). Because IoT is known for ubiquitous computing (using embedded sensors and actuators) and pervasive sensing technologies (such as RFID tags), it is capable of automating manufacturing processes by connecting humans, machines, and manufacturing processes [8].

Cloud manufacturing is emerging as a new manufacturing paradigm which applies well-known basic concepts from cloud computing to manufacturing processes and deliver shared, ubiquitous, on-demand manufacturing services. It is an innovative, web-based manufacturing model. It is promising to transform today's manufacturing industry from production-oriented to service-oriented, highly collaborative manufacturing of the future.

Cloud manufacturing is a service-oriented, knowledge-based smart manufacturing system. It is integrated with cloud computing technology, Internet of Things, and high-performance computing. Through cloud manufacturing, manufacturing resources *(such as manufacturing software tools, manufacturing equipment, and manufacturing capabilities)* are "virtualized" and offered as consumable in the same way as electricity, gas, and water. The manufacturing resources are provided to users as services over the Internet (cloud) in a pay-as-you-go manner [11, 12].

Cloud manufacturing is inspired by cloud computing. Cloud computing enables manufacturers to use new forms of production systems such as 3D printing (additive manufacturing), industrial robots, and high-performance computing. Just as cloud computer share pool of computing resources, cloud manufacturing enables on-demand network access to a shared poor of manufacturing resources such as manufacturing equipment, software, materials, storage, assets, and capabilities. CM is regarded as Manufacturing-as-a-Service (MaaS) and is expected to help the

Fig. 13.2 Cloud manufacturing architecture [7]

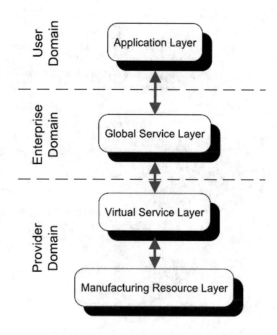

enterprise achieve the business benefits of reducing the idle capacity and increasing utilization [13, 14].

There are two types of cloud computing adoptions in manufacturing: adapting cloud computing technologies directly to manufacturing or using the manufacturing version of cloud computing, which is known as cloud manufacturing.

As shown in Fig. 13.2, the cloud manufacturing architecture is composed of the following four layers [7, 9]:

1. *Manufacturing resource layer*: This includes the physical resources (e.g., hardware; software) and capabilities.
2. *Virtual service layer*: This identifies and virtualizes manufacturing resources and capabilities and packages them on the cloud as services. Supporting technologies include RFID, wireless sensor networks, industrial Internet, cyber physical systems, and cloud computing.
3. *Global service layer*: This controls the entire operational activities of the cloud and enables effective interconnection between physical and cyber objects through industrial Internet.
4. *Application layer*: Provides an interface between the user and cloud services.

Cloud manufacturing plays a crucial role in many manufacturing fields such as the mold manufacturing, 3D printing, semiconductor manufacturing, sheet metal industry, machining, robotics, and product development. Figure 13.3 shows a typical system of cloud manufacturing [15].

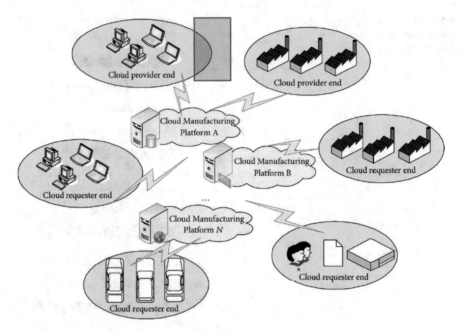

Fig. 13.3 The system of cloud manufacturing [15]

13.4   Migrating to the Cloud

Any company in the manufacturing business must fight to stay relevant and competitive.

Manufacturing companies of all sizes can benefit greatly from cloud computing. Companies small and big must realize that the cloud offers them a way to improve operations and reduce time. A manufacturer can migrate to the cloud by taking the following three key steps [16]:

1. *Define your business objectives:* Create clarity around how, exactly, you expect cloud solutions to maximize business efficiency. The idea here is that cloud should not simply provide a new way of provisioning IT resources—but enable better ways to operate. You may need an effective strategic partner for your cloud migration. It is recommended to find a partner that has all the futures that are important to your business.
2. *Clarify your IT expectations*: Next, clearly define which kinds of IT benefits to optimize for. Your list might include items like reduced IT infrastructure footprints at manufacturing sites, better price-to-performance ratios due to the scaling flexibility of cloud solutions, better disaster-recovery capabilities, and automation.
3. *Prioritize cloud security:* Finally, shape your cloud security strategy, governance, and infrastructure. Given how your cloud solutions will connect to business-critical IT and operational technology infrastructure, having these in place is a must.

13.5   Applications

Manufacturing is a critical component of many businesses. Here are some of the sectors where companies are already utilizing cloud manufacturing [17]:

- Automobile
- Aerospace
- Electronics
- Retail
- Pharmaceutical
- Food and beverage
- Distribution
- Furniture

There are several ways the cloud can be used in manufacturing. Some common applications of cloud manufacturing include:

- *Automation:* This can have a huge benefit on your company. Its importance to the manufacturing industry will continue to skyrocket. CM quickens deployments and the production by automating communication between manufacturing and accounting. For example, customer services is as important in manufacturing as it is any other business. The cloud offers technologies for automating customer service, taking orders online, and allowing users to check the status of orders at any time. As automation adaption increases, automated equipment and process monitoring become important.
- *Machining Industry:* The machine-tool manufacturing enterprises use cloud manufacturing to develop intelligent machine tools and novel business models to increase their competitiveness and gain profits [18].
- *Cloud Robotics:* The availability of new technologies, such as computing cloud and big data, and their possible uses in robotics have led to a new line of research called *cloud robotics*, which are used for grasping and navigation. The use of cloud robotics is based on the integration of cloud computing and industrial robots. Storing large amounts of data in the cloud is a key solution addressing robot limitations [19, 20].
- *Electronic Industry*: The concept of cloud manufacturing has been applied to the semiconductor manufacturing operations such as to wafer fab simulation, chain management, test operations management [21]. With cloud manufacturing, PCB prototyping should be faster.
- *Cloud-Based Marketing:* The comprehensive nature of cloud technology makes it suitable for marketing campaigns. Manufacturers use cloud-based applications to aid in planning, executing, and managing marketing campaigns and also to track the effectiveness of marketing campaigns [22].
- *Product Development:* Cloud manufacturing software is increasingly sought out by companies with heavy production components because it serves as a tool that assists companies in streamlining and visualizing the production process from creation to distribution. As a cloud-based solution, these systems provide mobile support that allows employees in the field or on the production floor to access important information at various locations anytime and anywhere [23].

13.6   Benefits

For manufacturers, the cloud provides endless possibilities for networking and sharing resources. The cloud manufacturing is becoming popular with manufacturers because it caters to their needs and provides services at lower cost. It is helping businesses redefine their niche and rebrand themselves so they can attain new manufacturing heights. It is an excellent choice for companies with small IT departments and ideal for companies with no IT. The IT team does not have to deal with the upkeep of the software. This results in reduction in capital expenditure. Using cloud-based systems to streamline key areas of their business, manufacturers are freeing up more time to invest in new products and selling more [24]. It supports the needs of small businesses by making small quantity runs affordable. It enables an organization or a business to find a manufacturing service partner that offers real-time quotes online.

Other benefits of cloud manufacturers include the following [25–29]:

- *Sustainability:* Sustainability should be at the top of the list of considerations of manufacturers. The major reason for the growing popularity in cloud manufacturing is its sustainability. Manufacturers should change the ways they implement operations, use data, engage with customers, and create an overall more sustainable ecosystem. This can be achieved with a cloud-based strategy.
- *Scalability*: Cloud computing can grow with the business or can easily scale back during slower times.
- *Reliability:* Cloud-based software technology has fewer technical problems than software used on individual computers.
- *Cost-Effectiveness:* Cloud solutions do not require in-house servers. Cost is one of the leading factors when manufacturers choose cloud-based solutions for enhancements to their ERP software. The cloud can make a difference in minimizing costs in manufacturing operations.
- *Short Go-To-Market Times:* Cloud manufacturing clearly reduces the go-to-market times. Cloud solutions improve turnaround time and reduce time to market for enhancement of market share.
- *Marketplace Advantage Over Companies Not Using Cloud Technology:* Better productivity and communications give companies that use cloud manufacturing solutions a competitive edge.
- *Centralized Management:* Access programs can be deployed from any computer in the organization, improving management capabilities.
- *Less Points of Failure:* Cloud manufacturing encompasses all steps from quoting to delivery. Therefore, there is only one point of contact and less points of possible failure.
- *Updated Equipment*: Technology based in the cloud stays updated without IT departments wasting time to ensure that everyone has the latest version of the software. Cloud-based solutions eliminate the delays, errors, and frustrations common in the traditional email-and-wait experience. Users or manufacturers do not need to invest on high-end computers or worry about software upgrades or

updates. The cloud eliminates the challenge of regularly updating equipment and machinery as one can upgrade them without a huge upfront cost. Instead, one pays a monthly fee to access the necessary resources.

- *Minimal Environmental Impact:* Cloud manufacturing can significantly reduce the environmental impact of making and moving goods around the world. It enables customers to make choices about designs and materials that can lead to more sustainable parts.
- *Prototyping*: Prototyping is a traditionally difficult part of the product development process. Cloud manufacturing is designed to make it easy, inexpensive, and quick to produce prototypes and small batches. This allows designers to iterate, improve, and deliver on time.
- *Location Irrelevance:* The cloud has introduced the notion of "location irrelevance" to manufacturing. According to this concept, it should not matter where the product stays, whether you make it, or somebody makes it for you. This is especially true for modern manufacturing where product planners, designers, and manufacturers who collaborate do not need to be at the same specific location at the same time.

Figure 13.4 shows some of the benefits of cloud manufacturing.

Fig. 13.4 Some benefits of cloud manufacturing

13.7   Challenges

There are some inherent drawbacks in adopting cloud-based applications. CM is in its infancy, and there are lot of problems that need to be resolved for it to become mature.

There is lack of software frameworks to support cloud manufacturing. Other challenges include the following [4, 30–32]:

- *Security:* Data security is a key concern. The main challenge involves the security for mission critical applications. Manufacturers need to be able to share information without exposing sensitive and confidential materials such as their intellectual property. Trust and reputation play crucial roles in the service-oriented network manufacturing paradigm.
- *Data Breaches:* With manufacturing emerging as a hot industry, data breaches are common. Cybercriminals both within and outside a company are waiting for opportune moments to exfiltrate sensitive data for espionage and monetary gains. Manufacturing's most commonly compromised data includes credentials, internal operations data, confidential patents and invention disclosures, and other company secrets.
- *Continuing Costs:* Although the subscription cost of a cloud-based system may be an advantage in most cases, it is observed that a cloud-based solution has no upfront investment; nonetheless, there is an ongoing monthly fee to use the service. Regular updates could bring on additional costs.
- *Lack of Standard*: To enable seamless switching between factories, tooling and materials must be standardized to be readily consumed by the given factory. The manufacturing industry lacks a uniform data standard because of the wide variety of manufacturing resources and business demand, supply, spare-parts variations, and order volatility. There is an added challenge of a missing standardized communication protocol across different machinery equipment from different providers.
- *Human Factors:* These can have a pivotal role in enabling the adoption of CM, while ensuring the safety and well-being of users involved in a CM environment.

These challenges are illustrated in Fig. 13.5. These are no small concerns; rather they are being rapidly addressed because most companies are not ready to run high-speed manufacturing productions systems in the cloud. To address these concerns, some manufacturers are implementing internal/private clouds or hybrid public/private clouds.

13.8   Global Cloud Manufacturing

Cloud manufacturing (CM) has emerged as a service-oriented paradigm that enables modularization and on-demand servitization or the on-demand and complete reinventing of the manufacturing business model and its associated resources. In order

Fig. 13.5 Cloud manufacturing challenges

for manufacturers worldwide to remain relevant and competitive in the highly dynamic new environment, they need to look toward digital transformation, specifically with cloud manufacturing. Every business should be regarded as being on a digital journey. North America has the maximum global cloud manufacturing market coverage as several market leaders belong to this zone. We now consider how some nations adopt cloud manufacturing.

- *United States:* The US National Institute of Standards and Technology (NIST) defines cloud computing (CC) as "a model for enabling ubiquitous, convenient, on-demand network access to a shared pool of configurable computing resources (e.g., networks, servers, storage, applications, and services) that can be rapidly provisioned and released with minimal management effort or service provider interaction." Cloud computing is emerging as one of the major enablers for the manufacturing industry. Building on NIST's definition of CC, CM is regarded as a computing and service-oriented manufacturing model developed from existing advanced manufacturing models [33]. Most of US manufacturers are reluctant to adopt cloud manufacturing. The state of American manufacturing has become debatable. On the one hand, manufacturing is seen as a struggling sector of the economy. Twenty years ago, US manufacturing employed approximately 17 million Americans. However, today, that number has fallen to just 12 million, a loss of five million jobs. Cloud computing is changing the manufacturing paradigm for large and small manufacturers. It is innovating, reducing cost, and bolstering the competitiveness of American manufacturing. For example, automakers such as Ford, Tesla, and Hyundai (among others) are using the cloud to deliver over-the-air updates to vehicles' powertrain, infotainment, navigation, and safety systems. Ford is piloting a cloud-based solution on Google Earth to enable virtual navigation of its assembly plants. Drug-maker Pfizer has leveraged cloud computing to completely re-engineer its complex global supply chain. Merck has successfully leveraged cloud computing to improve its manufacturing process for vaccines [34].

- *United Kingdom*: UK economic prosperity increasingly depends on developing and maintaining a resilient and sustainable manufacturing sector. The sector must be capable of producing a variety of products faster, better, and more affordable. It should also minimize the use of materials, energy, transport, and resources while maximizing environmental sustainability and economic competitiveness [5].
- *China:* The manufacturing industry is an important pillar of China's economy. The Chinese government regards manufacturing technology as a strategic measure for realizing modernization and industrialization. Advanced manufacturing tools can be used to improve product quality, service, testing, training, usage, maintenance, and energy efficiency. Production costs and development times are also reduced. Service, resource environment, and knowledge-based innovation are key factors for enhancing core competitiveness in the manufacturing industry. In 2009, China became the world's second largest industrialized manufacturing country, second only to the United States. However, it has not yet turned into a high-end manufacturing power. "Made in China" is generally considered to be at the low end of the international value chain. The concept of cloud manufacturing was proposed for promoting the development of manufacturing informatization. Research into cloud manufacturing technology will accelerate the development of an intelligently networked, service-oriented manufacturing industry in China [35, 36].

13.9   Future of Cloud Manufacturing

Cloud manufacturing enable manufacturing organization to manage their portfolio of products using the same tools software companies have been leveraging for years. Cloud manufacturing is the next great advance in lean production. Cloud-based manufacturing will be the industry's future because it represents the future of how things will be fabricated. That is good for both manufacturers and consumers. Cloud manufacturing is poised to become the new service-oriented manufacturing mode. It is the future of manufacturing.

The future development of cloud manufacturing will face some challenges. Besides the integration technologies of cloud computing, Internet of Things, and embedded systems, several important technical issues must be solved such as cloud management engines, collaboration between cloud manufacturing applications, and visualization and user interfaces in cloud environments.

As of now, most companies are not ready to run high-speed manufacturing production systems in the cloud, but there should be no doubt that cloud technology is coming into manufacturing [37]. Now is the right time for manufacturers to embrace the cloud. Manufacturing leaders must remember that not evolving and not embracing change means to be out of business.

13.10 Conclusion

Cloud manufacturing is a new manufacturing model that is service oriented, knowledge based, high performance, and energy efficient. It is a service-oriented, data-centric, and demand-driven business model to share manufacturing capabilities and resources on a cloud platform. The concept of cloud manufacturing has started to impact all walks of life (government, industry, academia, research). CM is transforming the traditional manufacturing business model, while helping the manufacturer to create intelligent factory networks that enable collaboration across the whole enterprise. Most companies are just scratching the surface of realizing their data's potential. Today, manufacturers face the choice to embrace cloud technology or risk losing out on competitiveness in the ever-evolving marketplace.

Although cloud manufacturing is still in the infant stage of development, it is gaining significant attention from academia, industry, and government. Every business should be embraced by the cloud manufacturing granted the immense opportunities it offers for stimulating rapid industrial growth, sustained job creation, and enhanced profits. More information on cloud manufacturing can be found in the books in [36–38] and the following related journals:

- *International Journal of Manufacturing Research.*
- *Manufacturing Letters.*
- *Journal of Manufacturing Systems.*
- *Journal of Advanced Manufacturing Technology.*
- *International Journal of Computer Integrated Manufacturing.*
- *International Journal of Advanced Manufacturing Technology.*

References

1. O. Fisher et al., Cloud manufacturing as a sustainable process manufacturing route. J. Manuf. Syst. **47**, 53–68 (2018)
2. L. Ren et al., Cloud manufacturing: From concept to practice. Enterp Inf Syst **9**(2), 186–209 (2015)
3. Cloud manufacturing, *Wikipedia*, the free encyclopedia https://en.wikipedia.org/wiki/Cloud_manufacturing
4. S. Ezell, B. Swanson, How cloud computing enables modern manufacturing (2017, June), https://www2.itif.org/2017-cloud-computing-enables-manufacturing.pdf

5. L. Xiao-gang, M. Jian, The overview of cloud manufacturing technology research. Int. J. Res. Eng. Sci. **3**(8), 7–14 (2015, August)
6. M.N.O. Sadiku, S.M. Musa, O.D. Momoh, Cloud computing: Opportunities and challenges. IEEE Potentials **33**(1), 34–36 (2014, January/February)
7. X. Xu, From cloud computing to cloud manufacturing. Robot. Comput. Integr. Manuf. **28**, 75–86 (2012)
8. D. Wu et al., Cloud-based design and manufacturing: A new paradigm in digital manufacturing and design innovation. Comput. Aided Des. **59**, 1–14 (2015, February)
9. M. Moghaddam, J.R. Silva, S.Y. Nof, Manufacturing-as-a-service—From e-work and service-oriented architecture to the cloud manufacturing paradigm. IFAC-Pap OnLine **48**(3), 828–833 (2015)
10. M.N.O. Sadiku, *Emerging Internet-Based Technologies* (CRC Press, Boca Raton, FL, 2019), pp. 63–79
11. M.N.O. Sadiku, Y. Wang, S. Cui, S.M. Musa, Cloud manufacturing. Int. J. Adv. Res. Comput. Sci. Softw. Eng. **8**(2), 39–41 (2018, February)
12. M.N.O. Sadiku, T.J. Ashaolu, A. Ajayi-Majebi, S.M. Musa, The essence of cloud manufacturing. Int. J. Sci. Adv. **1**(2) (2020, September–October)
13. L. Wang, An overview of internet-enabled cloud-based cyber manufacturing. Trans. Inst. Meas. Control. **39**(4), 388–397 (2017)
14. L. Ren et al., Cloud manufacturing: Key characteristics and applications. Int. J. Comput. Integr. Manuf. **30**(6), 501–515 (2017)
15. L. Zhu, Y. Zhao, W. Wang, A bilayer resource model for cloud manufacturing services. Math. Probl. Eng. (2013)
16. Cloud-enabled manufacturing (2021, January), https://www.accenture.com/us-en/insights/industry-x/cloud-enabled-manufacturing
17. L. Jenkins, The best cloud manufacturing ERP software, https://www.selecthub.com/enterprise-resource-planning/myths-cloud-erp-manufacturing/
18. C.C. Chen et al., A novel cloud manufacturing framework with auto-scaling capability for the machining industry. Int. J. Comput. Integr. Manuf. **29**(7), 786–804 (2016)
19. X.V. Wang et al., Ubiquitous manufacturing system based on cloud: A robotics application. Robot. Comput. Integr. Manuf. **45**, 116–125 (2017)
20. H. Yan et al., Cloud robotics in smart manufacturing environments: Challenges and countermeasures. Comput. Electr. Eng. **63**, 56–65 (2017)
21. X. Wu, F. Qiao, K. Poon, Cloud manufacturing application in semiconductor industry, in *Proceedings of the 2014 Winter Simulation Conference,* 2014, pp. 2376–2383
22. How the manufacturing industry uses cloud technology (2020, October), https://mantec.org/how-the-manufacturing-industry-uses-cloud-tech/
23. A. Velling, What is cloud manufacturing? And who can benefit from it? (2019, October), https://fractory.com/cloud-manufacturing/
24. D. Emerson, Cloud computing's effect on manufacturing, https://www.areadevelopment.com/advanced-manufacturing/December-2015/Cloud-Computing-Effect-on-Manufacturing-234455.shtml
25. C. First, What is cloud manufacturing? (2016, June), http://erpsoftwareblog.com/cloud/2016/06/what-is-cloud-manufacturing/
26. 10 ways cloud computing is revolutionizing manufacturing, https://www.forbes.com/sites/louiscolumbus/2013/05/06/ten-ways-cloud-computing-is-revolutionizing-manufacturing/#20961d58859c
27. S. Saslow, How cloud-based manufacturing applications can revolutionize your plant (2014, October), https://itgcloud.com/how-cloud-based-manufacturing-applications-can-revolutionize-your-plant/
28. D. Golightly et al., Manufacturing in the cloud: A human factors perspective, vol. 55, 2016, September, pp 12–21

29. L. Columbus, The importance of securing multi-cloud manufacturing systems in a zero trust world (2019, November), https://cloudcomputing-news.net/news/2019/nov/14/securing-multi-cloud-manufacturing-systems-in-a-zero-trust-world/
30. Advantages and disadvantages of cloud-based manufacturing software (2020, January), https://manufacturing-software-blog.mrpeasy.com/advantages-and-disadvantages-of-cloud-based-manufacturing-software/
31. Y. Yadekar, E. Shehab, J. Mehnen, Challenges of cloud technology in manufacturing environment, in *Proceedings of the 11th International Conference on Manufacturing Research*, UK, September 2013, pp 177–182
32. L. Zeballos, O. Quiroga, From computer integrated manufacturing to cloud manufacturing, http://www.clei2017-46jaiio.sadio.org.ar/sites/default/files/Mem/SII/sii-16.pdf
33. Advanced manufacturing technology research group, https://www.nottingham.ac.uk/research/groups/advanced-manufacturing-technology-research-group/research/digital-manufacturing/cloud-manufacturing.aspx
34. L. Bohu, Z. Lin, C. Xudong, Introduction to cloud manufacturing (2019, December), https://www.zte.com.cn/global/about/magazine/zte-communications/2010/4/en_140/196752.html
35. M. Davidson, The future of cloud-based manufacturing applications (2013, July), https://blog.lnsresearch.com/blog/bid/183519/The-Future-of-Cloud-Based-Manufacturing-Applications
36. W. Li, J. Mehnen (eds.), *Cloud Manufacturing: Distributed Computing Technologies for Global and Sustainable Manufacturing* (Springer, 2013)
37. D. Schaefer (ed.), *Cloud-Based Design and Manufacturing (CBDM):A Service-Oriented Product Development Paradigm for the 21st Century* (Springer, 2014)
38. B. Sokolov, D. Ivanov, A. Dolgui (eds.), *Scheduling in Industry 4.0 and Cloud Manufacturing* (Springer, 2020)

Chapter 14
Nanomanufacturing

We believe that if this technology has a future, we will need to have a large research community working on it.

–Despont

14.1 Introduction

The nanoworld is a mysterious place because materials behave differently at the atomic or nanoscale. The discovery of higher resolution microscopes in the 1980s enabled having insight into nanoscale material structures and their properties. Nanotechnology (science on the scale of single atoms and molecules) has been identified as the "fourth industrial revolution" to disrupt the modern world. It is so described because of the special properties of materials at the nanoscale. It has permeated all sectors of our economy due to the unique properties of materials at the nanoscale. It is transforming the world of materials and its influence will be broad. It will not only initiate the next industrial revolution, but it will also offer technological solutions.

Nanotechnology refers to the science of nanomaterials. It is the measuring, modeling, and manipulating of matter at atomic scale or in the dimension of 1–100 nanometers (nm). (A nanometer is 1 billionth of a meter). It offers the opportunity to produce new structures, materials, and devices with unique properties such as conductivity, strength, and chemical reactivity. Electrical and mechanical properties can change at the nanoscale. For example, at the nanoscale, gold becomes an active catalyst, helping to turn chemicals X and Y into product Z. Most nanomaterials are made by chemical processes, which may or may not generate pollutants or waste materials [1].

Manufacturing is the major industry for promoting economic and social development. It is a dynamic, ever-evolving industry, with newer and better technologies being developed on a regular basis. Nanomanufacturing is the manufacturing processes of objects or materials with dimensions between one and 100 nm. It is the scaled-up, cost-effective manufacturing of nanoscale materials, structures, devices, and systems. It focuses on developing scalable, high-yield processes for the production of materials, structures, devices, and systems at the nanoscale.

© The Author(s), under exclusive license to Springer Nature Switzerland AG 2023
M. N. O. Sadiku et al., *Emerging Technologies in Manufacturing*,
https://doi.org/10.1007/978-3-031-23156-8_14

This chapter provides an introduction to nanomanufacturing. It begins by giving an overview of nanotechnology and nanomanufacturing. It presents same applications of nanomanufacturing. It highlights some benefits and challenges of nanomanufacturing. It covers the essence of global nanomanufacturing and the future of nanomanufacturing. The last section concludes with comments.

14.2 Overview of Nanotechnology

We must first understand what nanotechnology is about before we can understand nanomanufacturing. Techniques are now available which make it possible to manipulate materials on the atomic or molecular scale to produce objects which are no more than a few nanometers in diameter. The processes used to make and manipulate such materials are known as *nanotechnology*; the materials or objects themselves are called *nanomaterials*, and the study and discovery of these materials are known as *nanoscience*. Thus, nanotechnology is basically the control of matter and processes at the nanoscale.

Nanotechnology involves the manipulation of matter at the atomic and molecular scales. It is emerging as a principal discipline that is integrating chemistry and material science. It holds the promise of being a main driver of technology with significant impact for all aspects of society. It is an interdisciplinary field covering physics, chemistry, biology, materials science, and engineering [2].

Richard Feynman, the Nobel Prize-winning physicist, introduced the world to nanotechnology in 1959. The term "nanotechnology" was coined in 1974 by Norio Taniguchi, a professor at Tokyo Science University. Nanotechnology involves the manipulation of atoms and molecules at the nanoscale so that materials have new unique properties. Nanotechnology is a multidisciplinary field that includes biology, chemistry, physics, material science, and engineering. It is the science of small things—at the atomic level or nanoscale level [3]. Nanotechnology also includes domains like nanoscience, nanomaterials, nanomedicine, nanomeasurement, nanomanipulation, nanoelectronics, and nanorobotics.

Nanotechnology is the science of small things—at the atomic level or nanoscale level. It has the idea that the technology of the future will be built on atoms. It has impact on every area of science and technology. Nanotechnology involves imaging, measuring, modeling, and manipulating matter at the nanoscale. At this level, the physical, chemical, and biological properties of materials fundamentally differ from the properties of individual atoms and molecules or bulk matter [4].

Nanotechnology covers a wide variety of disciplines like physics, chemistry, biology, biotechnology, information technology, engineering, and their potential applications.

Some of the sectors covered by nanotechnology are shown in Fig. 14.1 [5]. Nanotechnology holds great potential for pollution prevention and sustainability.

It has given manufacturers new and better ways to produce materials of varying strengths, weights, volumes, surface areas, lengths, and thicknesses.

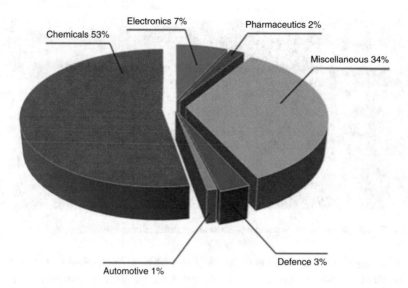

Electronics 7%   Pharmaceutics 2%

Chemicals 53%

Miscellaneous 34%

Automotive 1%   Defence 3%

Fig. 14.1 Some sectors covered by nanotechnology [5]

Widespread commercialization of nanotechnologies has been restricted by lack of the means to produce these technologies at scale. Therefore, the next major step facing nanotechnologists is figuring out how to move from laboratory discoveries to commercial products, i.e., how to manufacture devices with nanoscale features in a cost-effective manner.

14.3 Nanomanufacturing

Nanotechnology has only come into wide use in the manufacturing sector recently. A series of breakthroughs in materials and designs have allowed manufacturers to work at scales smaller than a billionth of a meter. Nanomanufacturing (or nanofabrication) is manufacturing at the nanoscale. The terms "nanofabrication" and "nanomanufacturing" are often used interchangeably. Nanomanufacturing (may be regarded as a set of industrial processes based on nanotechnology used in developing products at the nanoscale. It involves reliable and cost-effective manufacturing of nanoscale materials, structures, devices, and systems. This area is increasingly dynamic, with supporting advanced technologies that are vital when manufacturing parts or features at such small scales are emerging weekly. It is an interdisciplinary field covering physics, chemistry, biology, materials science, and engineering.

Some key features of nanomanufacturing are discussed as follows [6, 7]:

1. *Top-Down or Bottom-Up:* There are two basic approaches to nanomanufacturing: top-down and bottom-up. Top-down nanomanufacturing reduces large pieces of materials all the way down to the nanoscale. A manufacturer will start

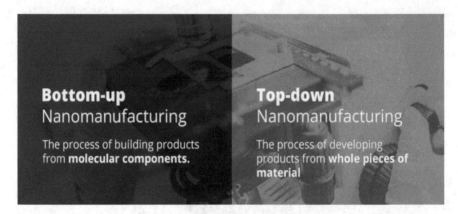

Fig. 14.2 Comparing bottom-up and top-down nanomanufacturing approaches [8]

with larger materials and use chemical and physical processes to break them down into nanoscopic elements. This approach requires larger amounts of materials and can lead to waste due to the need for discarding material. The bottom-up (or additive) approach to fabrication creates products by building them up from atomic- and molecular-scale components, which can be time-consuming. For example, carbon nanotubes are manufactured with a bottom-up approach. The top-down method is the more common approach, while the bottom-up method is still in the experimental stage of development. These two approaches are compared in Fig. 14.2 [8].

Various processes have been developed under the categories of bottom-up and top-down nanomanufacturing. Some of the more common processes include the following [8]:

- Chemical vapor deposition (CVD): This process is used to make materials from a series of chemical reactions. CVD is often employed in the semiconductor industry, where it is used to create film strips with silicone, carbon, and filaments.
- Molecular-beam epitaxy (MBE): A process used to deposit thin films.
- Atomic layer epitaxy (ALE): A process where layers comprised of single-atom thickness are deposited onto surfaces.
- Dip-pen nanolithography (DPN): A process that resembles the functions of an ink pen; DPN involves the use of a chemically saturated microscopic tip, which is used to write on surfaces. A lithograph that heats above 100 °C is capable of fabricating metal electrodes far more effectively than a conventional electron beam.
- Nanoimprint lithography (NIL): A nanoscale stamping process where features are imprinted onto a surface.
- Roll-to-roll processing: A process where ultrathin strips of metal and plastic are imprinted with nanoscale devices at a very high frequency.
- Self-assembly: A set of processes where different elements are brought together without process intervention, to form a structure.

2. *Nanomaterials in Manufacturing:* A nanomaterial (NM) (or nano-sized mate-rial) is the material with any external dimension in the nanoscale. Nanomaterials are basically chemical substances or materials that are manufactured and used at a very small scale.

For example, miniaturization enables microprocessors developed using these parts, to operate faster. However, there are a number of technical challenges to achieving these advancements, such as the lack of ultrafine precursors to make these parts, inadequate dissipation of the large amounts of heat generated by these micro-processors, and poor reliability. Nanomaterials help overcome these barriers by offering manufacturers materials with better thermal conductivity, nanocrystalline starting materials, ultra-high-purity materials, and materials with longer-lasting, and more durable interconnections. The use of nanoparticles in the manufacturing industry will continues to grow [9].

3. *Nanomanufacturing Methods:* Specialized equipment and techniques are neces-sary to manipulate matter at the nanoscale. There are many methods for this, including self-assembly, photolithography, and dip-pen lithography. Self-assembly means that a group of nanocomponents has to come together to form an organized structure. This self-assembly would occur from the bottom-up. Photolithography is analogous to the negative obtained with film, as it uses light exposure to project images. Dip-pen lithography relies on and enables the tip of an atomic force microscope to be imbued with chemicals that are subsequently used to create patterns.

Figure 14.3 shows a nanomanufacturing facility [10].

14.4 Applications

Nanomanufacturing is revolutionizing many manufacturing sectors, including information technology, defense, medicine, transportation, energy, environmental science, telecommunications, and electronics. Common applications of nanomanu-facturing include [11, 12]:

- *Automotive and Aerospace*: The manufacturing industry will experience huge developments in sectors like automotive and aerospace. Nanotechnology is used in the auto industry, to create batteries, fuel cells, and converters that yield higher efficiencies. Tire manufacturers are increasingly using polymer nanocomposites to increase the durability and wear resistance of tires. Since nanotechnology cre-ates smaller and lighter crafts, less fuel will be needed to pilot aircraft.
- *Healthcare:* Nanotechnologies have enabled engineers and clinicians to collabo-rate in solving complex problems which require advanced nanomanufacturing capabilities to develop medical applications. Introducing these technologies and disseminating these results to healthcare engineering will greatly benefit the majority of population in developing and under-developed countries that would

Fig. 14.3 Nanomanufacturing facility [10]

be enabled to receive appropriate and affordable medical care, to achieve improvements in their quality of life.

- *Nanomaterials:* Popular nanomaterials, like carbon nanotubes, are already widely fabricated and applied in the manufacture of a variety of goods, including sailboat hulls, bicycle frames, and spaceship components. Nanotechnology can also be used to create more effective and stable lubricants, which are useful in a variety of industrial applications.
- *Electronics*: In electronics, nanotechnology enables the manufacture of tiny electronics and electric devices, like nanoscale transistors made of carbon nanotubes. In electronics, design at the nanoscale is creating highly flexible devices and circuit boards. Today, the only industry where nanomanufacturing is employed on a large scale is the semiconductor industry. The feature sizes of semiconductor devices have shrunk.
- *Nanomachines:* Nanotechnology has shown serious in nanomachines or nanites—mechanical or robotic devices that operate at the nanoscale. Nanoscale robots (called nanomachines or nanites) may soon revolutionize medical device construction. Nanomachines are, for the most part, future technologies that are not widely used in manufacturing right now.
- *Additive Nanomanufacturing:* Additive manufacturing has provided a quick way to fabricate inexpensive specialized components and spare parts. It cost-effectively lowers manufacturing inputs and outputs in small batch production. It is used by engineers for rapid prototyping and low-volume production. At the

nanoscale, additive nanomanufacturing provides a future path for new nanotechnology applications. It empowers smaller facilities to design, create, and manufacture on their own [13].

- *Drug Delivery:* One of the major challenges in development of nano-based drug delivery is its large-scale production. This includes manufacturing nanodevices below 20 nm. Producing nano drugs on a large scale requires the integration of methods as well as transfer of technology for the production of nano drugs at a large scale. A properly designed scale-up production reduces time and improves cost-effectiveness [14].
- *Self-Assembly:* As miniaturization reaches the nanoscale, conventional manufacturing technologies fail. Self-assembly is essentially one strategy for organizing matter on larger scales. It has become an important concept in nanotechnology. The key to using self-assembly, as a controlled and directed fabrication process, lies in designing the components that are required to self-assemble into desired patterns and functions. Until robotic assemblers capable of nanofabrication can be built, self-assembly will be the necessary technology required to develop for bottom-up fabrication. For example, IBM announced their self-assembling structures in their airgap processors. Self-assembly is the reason that nanotechnologies have such an extensive influence on the chemical market [15].

Other applications of nanomanufacturing include laser nanomanufacturing, mobile computing, energy technologies batteries, solar panels, clothing, wearable devices and systems, coatings, lithium-ion, and sporting items such as baseball bats and tennis rackets (Fig. 14.4).

14.5 Benefits

Nanotechnology is regarded as the driving force behind a new industrial revolution. It presents opportunities to create new and better products. Many manufacturers are using nanotechnology to make products with improved capabilities or to reduce their manufacturing cost. Nanomanufacturing is flexible and is poised to meet customer needs and would support sustainable manufacturing as well as molecular manufacturing. It is widely regarded as critical to realizing practical applications of nanotechnology in areas such as nanoelectronics, nanobiotechnology, nanoscale structures, devices, and systems.

- *Small, but Mighty*: It is commonly said all things begin small, and nanotechnology is not an exception. Nanomanufacturing and nanotechnology deal with particles which are a hundred thousand times smaller than the width of a human hair.
- *Miniaturization:* Nanomanufacturing benefits from miniaturizing the components because scaling down emitters implies less power consumption, less bias voltage to operate them, and higher throughput.
- *Superior Performance:* Nanomanufacturing plays an important role for high performance products in several applications. Nanomaterials can provide superior

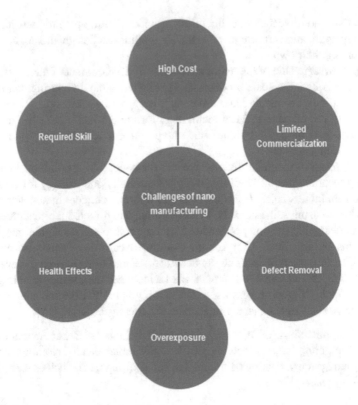

Fig. 14.4 Some challenges of nanomanufacturing

products in areas such as medicine, electronics, biomaterials, and consumer products.

- *Improve Efficiency*: Nanomanufacturing will improve efficiency in a number of operations, from design to packaging and to the transportation of goods.

14.6   Challenges

Nanomanufacturing is currently very much in its infancy. Nanomanufacturing presents unique challenges not inherent in traditional manufacturing. In many nanomanufacturing processes, the execution of many runs is not practical due to limited resources. It is challenging to maintain nanoscale properties and quality of nanosystem during high-rate and high-volume production as well as during the lifetime of the product after production. Nanomanufacturing is energy-intensive in terms of material processing and nano-feature manufacturing. It is currently difficult to assess the environmental, ethical, and social impacts of nanomanufacturing. Some other challenges of nanomanufacturing include [16]:

- *High Cost:* The commercial production of nanotechnology components often involves vast resources. For example, high-quality graphene is still very hard to manufacture in large quantities. Nanomanufacturing is dominated by lithography tools that are too expensive for small- and medium-sized enterprises (SMEs) to invest in.
- *Limited Commercialization*: Nanomanufacturing has been restricted by the lack of driving applications and the inability to produce nanotechnology devices at commercial scales, including mass production. Current nanomanufacturing processes may be wasteful and expensive if applied on a commercial scale. Scaling down assembly lines to the nano level is difficult.
- *Defect Removal:* Another challenge is testing reliability and establishing methods for defect control. Currently, defect control in the semiconductor industry is nonselective and takes 20–25% of the total manufacturing time. Removal of defects for nanoscale system is projected to take up much more time because it requires selective and careful removal of impurities.
- *Overexposure:* Harvesting nanomaterials from reactors results in high-risk exposures. Packing nanomaterials tends to release relatively large agglomerates of nanomaterials into the workplace. Protecting workers from health effects from overexposure to nanomaterials in the manufacturing environment is challenging.
- *Health Effects*: Concerns about the health effects of nanoparticles and nanofibers imply that calls for the tighter regulation of nanotechnology are growing. Based on some concerns regarding environmental, health, and safety (EHS) issues, significant research is needed to understand the risks associated with nanomaterials. Much remains unknown about EHS risks.
- *Required Skill:* Nanomanufacturing requires new skills that are needed for these high-tech manufacturing jobs. This may require specialized education, such as powder processing and metallurgical extraction.

There are some major challenges for nanomanufacturing to overcome before it can really start to fulfil its potential. The challenges are so complex that a multidisciplinary team of experts is needed to address them. The challenges may hamper attempts to transition nanotechnology from R&D to full-scale manufacturing. Even with the challenges, nanomanufacturing is still making inroads relative to advancement in nanotechnology.

14.7 Global Nanomanufacturing

Nanotechnology is constantly being researched and developed at laboratories all over the world. The promise of nanomanufacturing is such that a lot of money will be spent on research by national governments. We now consider how some nations pursue nanomanufacturing.

- *United States:* Many view the United States as the world's premier nanotechnology research and development (R&D) nation. The National Nanomanufacturing

Network (NNN) is an alliance of academic, government, and industry partners who are committed to advance nanomanufacturing in the United States. In 2000, the US government established the National Nanotechnology Initiative (NNI), as a nanotechnology R&D initiative. NNI was created with the goal of making the United States a global leader in nanotechnology. NNI partners with several centers across the United States to develop nanotechnology applications. Nanotechnology is revolutionizing many manufacturing sectors, including information technology, defense, medicine, transportation, energy, environmental science, telecommunications, and electronics.

Some nanotechnology companies in the United States include [17]:

– 10 Angstroms
– 3D Systems
– 3DIcon
– 3DM
– 4Wave
– A & A Company
– Abeam Technologies
– Accelergy
– Accium
– ACS Material
– Aculon
– Adámas Nanotechnologies
– ADA Technologies
– Advance Reproductions
– Advanced Ceramic Materials
– Advanced Diamond Technologies
– Advanced Energy Industries
– Advanced Micro Devices (AMD)
– Advanced Nano-Coatings
– Advanced Optical Technologies
– American Elements
– Antibodies Incorporated
– Berkeley Advanced Biomaterials
– Carbon Solutions, Inc.
– Chemat Technology
– Cospheric Nano
– Green Millennium

• *European Union:* In Europe, the manufacturing industry is facing a growing pressure derived from competitive economies. The number of consumer products claiming to contain nanomaterials doubles every 18 months. The European Society for Precision Engineering and Nanotechnology (euspen) is an influential community linking industrialists, researchers, and established players worldwide. Its mission is to advance the arts, sciences, and technology of nanotechnology, to promote its dissemination through education and training, and to facilitate its exploitation by science and industry. Euspen (www.euspen.eu) provides an entrepreneurial platform that enables companies and research institutions to promote their latest technology developments, products, and services [18]. It has been hosting conferences and events in nanomanufacturing for almost 20 years, bringing together leading experts from industry, R&D, and academia.

- *Australia:* Australia and Victoria have an active and advanced industry based on the manufacture of small particles. Australian manufacturers are the global pioneers in a number of advanced manufacturing areas, including advanced nanomaterial applications. This has made Australia to be the largest export earner in the global manufacturing industry. For example, advanced manufacturing accounted for $21.1 billion (18.3%) of all Australian export in 2001. Research institutions and industries across Australia have created nanocoating technologies that can be used in a wide range of applications [19].
- *United Kingdom: The* Nanomanufacturing Hub was established at the University of Surrey in UK with the purpose of fostering new collaborative developments in nanoscience and nanotechnology, from materials and characterization to devices and applications. The Hub will make the manufacturing of smart materials and self-powered electronics a reality in the United Kingdom. It brings together researchers with an international outlook in advanced materials and technologies such as nanotechnology. It will support small and medium-sized enterprises through the lower technology readiness levels (TRL) that require more research input and experience. A team of researchers are needed to cover the range of materials science, device design and physics, device fabrication, characterization, and testing [20]. Smart Nano NI is a Northern Ireland consortium involving local industry leading organizations working together on smart manufacturing. The consortium will provide opportunities for knowledge transfer to high-growth sectors across the whole United Kingdom [21].

14.8 Future of Nanomanufacturing

Nanotechnology has come to be applied to a wide range of small-scale engineering. It promises almost an indefinite range of applications in medicine, optics, communications, electronics, space, etc. The concept of manufacturing at nanoscale covers a host of different materials, devices, products, and processes. Nanoscale manufacturing is affecting many sectors of the economy and is having widely transformative impacts. It has the potential to set the pace for improvements across a wide range of industrial sectors. It could open new world markets.

The future of nanomanufacturing is bright, but progress will not come cheap. The future will be dictated by the trend of decreasing component sizes, material usages, and energy consumption of products. Resourceful R&D will need to rely on funding to further the potential of nanomanufacturing.

The global market for nanomaterials and nano-enabled products seems to be growing fast. To get a share of the market, a company must create and implement products fabricated by nanotechnology. Small companies looking for opportunities to try nanomanufacturing should find ways to embrace the revolution before it becomes mainstream.

14.9   Conclusion

Nanomanufacturing is the essential bridge between the discoveries of the nanosciences and real-world nanotechnology products. It embraces the processes and techniques utilized to produce nano-enabled products. It consists of a series of fabrication processes to build nanoscale structures and devices using nanotechnology. It is a relatively new branch of manufacturing that focuses on developing scalable, high-yield processes for the production of materials, structures, devices, and systems at the nanoscale.

Government agencies around the world are investing heavily in nanomanufacturing research and development. These governments provide researchers with the facilities, equipment, and trained workers to develop nanotechnology applications and associated manufacturing processes. More information about nanomanufacturing can be found in the books in [22–34] and the following related journals:

- *Nanotechnology.*
- *Nanoscale.*
- *Journal of Nanotechnology.*
- *Journal of Nanoscience and Nanotechnology,*
- *Journal of Micro and Nano-Manufacturing.*
- *Journal of Nanoengineering and Nanomanufacturing.*
- *ASME Journal of Micro- and Nano-Manufacturing*

References

1. M.N.O. Sadiku, *Emerging Green Technologies* (CRC Press., chapter 14, Boca Raton, FL, 2020), pp. 145–157
2. Ten things you should know about nanotechnology, https://www.nanowerk.com/nanotechnology/ten_things_you_should_know_5.php
3. M.N.O. Sadiku, M. Tembely, S.M. Musa, Nanotechnology: An introduction. Int. J. Softw. Hardw. Res. Eng. **4**(5), 40–44 (2016, May)
4. E.D. Sherly, K. Madgular, R. Kakkar, Green nanotechnology for cleaner environment present and future research needs. Curr. World Environ. **6**(1), 177–181 (2011)
5. O. Figovsky, D. Beilin, *Green Nanotechnology* (Pan Stanford Publishing, Singapore, 2017), p. xv
6. Manufacturing at the nanoscale, https://www.nano.gov/nanotech-101/what/manufacturing

7. Nanotechnology and manufacturing: The future is bright, https://www.gray.com/insights/nanotechnology-and-manufacturing-the-future-is-bright/

8. What is nanomanufacturing?, https://gesrepair.com/what-is-nanomanufacturing/

9. M.N.O. Sadiku, O.B. Egunjobi, S.M. Musa, Nano manufacturing: An introduction. Int. J. Trend Sci. Res. Dev. **5**(2), 1169–1172 (2021, Jan–Feb)

10. Nano manufacturing: Banknotes from butterflies, https://www.bloomberg.com/news/photo-essays/2015-02-25/nano-manufacturing-banknotes-from-butterflies

11. S. Huntington, Nanotechnology in manufacturing (2020, March), https://www.manufacturing-tomorrow.com/article/2020/03/nanotechnology-in-manufacturing/14945/

12. M. S. Packianather et al., Advanced micro and nano manufacturing technologies used in medical domain, in *Proceedings of the 6th International Conference on the Development of Biomedical Engineering in Vietnam*, Singapore, 2017, pp. 637–642

13. D.S. Engstrom et al., Additive nanomanufacturing – A review. J. Mater. Res. **29**, 1792–1816 (2014)

14. Nanomanufacturing systems, https://www.sciencedirect.com/topics/engineering/nanomanufacturing-system

15. Self-assembly, https://www.sciencedirect.com/topics/materials-science/self-assembly

16. Nano manufacturing, *Wikipedia,* the free encyclopedia, https://en.wikipedia.org/wiki/Nano_manufacturing

17. Nanotechnology companies in the USA, https://www.nanowerk.com/nanotechnology/Nanotechnology_Companies_in_the_USA.php

18. euspen hosts event on micro & nano manufacturing, https://www.euspen.eu/euspen-hosts-event-on-micro-nano-manufacturing/

19. Nanotechnology in the Australian advanced manufacturing industry, https://www.azonano.com/article.aspx?ArticleID=1631

20. Nanomanufacturing Hub, https://www.surrey.ac.uk/advanced-technology-institute/facilities/nanomanufacturing-hub

21. R. Fitton, Cirdan joins local industry leaders in Smart Nano NI consortium led by Seagate Technology in a bid to win major funding for Northern Ireland, 2021, February., https://www.cirdan.com/cirdan-joins-local-industry-leaders-in-smart-nano-ni-consortium-led-by-seagate-technology-in-a-bid-to-win-major-funding-for-northern-ireland/

22. M.J. Jackson (ed.), *Microfabrication and Nanomanufacturing* (CRC Press, Boca Raton, FL, 2005)

23. G. Tosello, H.N. Hansen (eds.), *Micro/Nano Manufacturing* (MDPI AG, 2017)

24. R. Gronheid, P. Nealey (eds.), *Directed Self-Assembly of Block Co-polymers for Nano-manufacturing* (Woodhead Publishing, Cambridge, 2015)

25. M.S. Shunmugam, M. Kanthababu (eds.), *Advances in Micro and Nano Manufacturing and Surface Engineering: Proceedings of AIMTDR 2018* (Springer, Singapore, 2019)

26. A. Hu (ed.), *Laser Micro-Nano-manufacturing and 3D Microprinting* (Springer, Cham, 2020)

27. A. Busnaina (ed.), *Nanomanufacturing Handbook* (CRC Press, Boca Raton, FL, 2007)

28. M.J. Jackson, *Micro-and Nanomanufacturing* (Springer, 2017)

29. W. Gao, *Precision Nanometrology: Sensors and Measuring Systems for Nanomanufacturing* (Springer, London, 2010)

30. H.N. Hansen, G. Tosello (eds.), *Micro/Nano Manufacturing* (MDPI Books, 2017)

31. R.M. Mahamood, E.T. Akinlabi, *Advanced Noncontact Cutting and Joining Technologies: Micro- and Nano-manufacturing* (Springer, Cham, 2018)

32. S.N. Joshi, P.L. Chandra (eds.), *Advanced Micro- and Nano-manufacturing Technologies Applications in Biochemical and Biomedical Engineering – Materials Horizons: From Nature to Nanomaterials* (Springer, Singapore, 2022)

33. B.D. McReynolds (ed.), *Nanomanufacturing: An Emerging Megatrend and Implications for the United States* (Nova Science Publishers, New York, 2014)

34. Y. Qin, *Micromanufacturing Engineering and Technology* (Elsevier, Oxford, 2010)

Chapter 15
Ubiquitous Manufacturing

*Data is the kind of ubiquitous resource that we can shape to
provide new innovations and new insights, and it's all around
us, and it can be mined very easily.*

— David McCandless

15.1 Introduction

Manufacturing encompasses the industrial production processes, through which
raw materials are transformed into finished products, which can be sold in the market to earn profits. The manufacturing gains could subsequently support organizational continuous improvement initiatives. Manufacturing is crucial to the national
economy; it is the key element of industrial competitiveness. Manufacturing companies are currently confronted with the challenges created by rapid technological
changes. Today's manufacturing enterprises have to deal with: (1) coverage of the
whole product life cycle, (2) environmentally conscious manufacturing, (3) simul-
taneous improvement of productivity and sustainability, and (4) enhancement of the
total performance of the product through its entire life cycle. Other challenges of
current manufacturing organizations include time-to-market reduction, mass customization realization, tougher multinational competition, flexibility, reconfigurability requirements, and speed/quality/cost improvements in product development
and manufacturing [1]. Thus, modern manufacturing requires a new generation of
production systems with better interoperability and new business models. With the
recent development of network technology, communication technology, sensor network technology, and computing technology, manufacturing resources, services,
facilities, and products can now be deployed ubiquitously.

The demand for highly customized products and the increasing popularity of
outsourcing has given birth to the concept of ubiquitous manufacturing (UM).
Ubiquitous manufacturing (UM or UbiM) is an application of ubiquitous computing
(or pervasive computing), which is a concept where computing is made to appear
everywhere and anywhere. UM emphasizes the mobility and dispersion of manufacturing resources, products, and users [2]. It may be regarded as a "design anywhere, make anywhere, sell anywhere, and at any time" paradigm.

© The Author(s), under exclusive license to Springer Nature Switzerland AG 2023 215
M. N. O. Sadiku et al., *Emerging Technologies in Manufacturing*,
https://doi.org/10.1007/978-3-031-23156-8_15

This chapter provides an introduction to ubiquitous manufacturing. It begins by providing some background on ubiquitous computing on which ubiquitous manufacturing is built. Then, it explains what ubiquitous manufacturing is all about. It describes technologies related to UM, the characteristics of UM, and its enabling technologies. It highlights the benefits and challenges of UM. It covers global ubiquitous manufacturing. The last section concludes with comments.

15.2   Overview on Ubiquitous Computing

Modern computing has made an impact on our daily life through the usage of communication technologies and the Internet. The need for using different devices that surround humans to communicate with each other at any time and any environment has introduced ubiquitous computing. Ubiquitous computing (ubicomp) is a concept in which computing is performed at any location. It is about computers everywhere surrounding humans, communicating with each other, and interacting with people and environments. It is the next wave in computing evolution. It seeks to embed computers into our everyday lives in such ways as to render them invisible and allow them to be taken for granted. It aims to revolutionize the current paradigm of human-computer interaction. It will have a profound effect on the way we interact with computers, devices, environment, and other people. Unlike desktop computing, ubiquitous computing can occur using any device, in any location, and in any format. Devices that use ubicomp have constant availability and are always connected [3, 4].

The word "ubiquitous" means existing everywhere. Ubiquitous computing (or pervasive computing) refers to an environment that enables people to use a variety of information and communication services by networking anywhere, anytime without any interruption. It makes computing power available at all times and all places in a convenient way. This is the idea behind the concept of ubiquitous manufacturing.

Ubiquitous computing is also known as pervasive computing. Ubicomp environment is marked by anywhere, anytime, and anyone-based characteristics. The terms "ubiquitous computing" and "pervasive computing" are used interchangeably, but they are conceptually different. Ubiquitous computing involves a three-step process: sense, compute, and actuate. It uses the advances in mobile computing and pervasive computing to present a global computing environment. Pervasive computing is invisible to human users; ubiquitous computing aims to provide pervasive computing environments to a human user as he moves from one place to another [5].

The term "ubiquitous computing" was coined by the late Mark Weiser around 1988 while at the Electronics and Imaging Laboratory of the Xerox Palo Alto Research Center. Mark Weiser is widely regarded as the father of ubiquitous computing. He had a vision of computing technology weaving itself into the very fabric of everyday life. He identified ubiquitous computing (many computers, one person) as the third generation of computing, following the first generation of mainframe

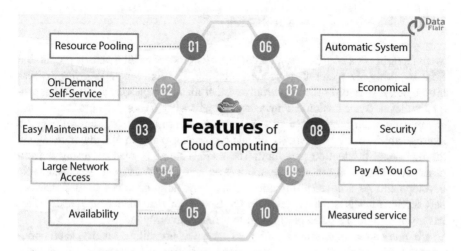

Fig. 15.1 Major characteristics of cloud computing [10]

computing (one computer, many people) and the second generation of personal computing (one computer, one person). The fourth generation of computing, known as collective computing, has emerged over the past decade [6–9].

Fig. 15.1 shows the major characteristics of cloud computing [10]. The basic features of ubiquitous computing include [11]:

- Computing elements are not integrated in a single workstation but distributed as everyday objects in user's work environment.
- The context-aware feature of ubiquitous environments is also called "situatedness." It requires adapting behavior based on information sensed from the physical and computational environment.
- Use of inexpensive processors, thereby reducing memory and storage requirements.
- Capturing real-time attributes.
- Totally connected and constantly available computing devices.
- Improvised and dynamic interactions among applications.
- Relies on converging the Internet, wireless technology, and advanced electronics.

The purpose of these features is to allow users to perform their tasks more effectively and efficiently. Ubiquitous computing is not just the omnipresence of computational systems; it represents new ways of thinking about human-machine interaction. We now live in a world where the number of computers exceeds the number of humans and are entering a many-to-one computer/human relationship. Ubiquitous computing truly is everywhere. From smartwatches to self-driving cars and everything in between. Ubiquitous computing is a paradigm, a lifestyle, and a technological way of thinking. This is one of the hallmarks that characterize the ubiquitous computing era [12].

15.3   Ubiquitous Manufacturing

Manufacturing is a coordinated and harmonized process of all levels of production from acquiring raw materials to product retailing. The entire manufacturing process is complex and requires the synchronization of many subsystems at a different production level. Since customers nowadays prefer highly customized products, modern manufacturing requires a new generation of production system with better interoperability and new business models. This leads to a need of a major change in the traditional production paradigm. The major change in manufacturing is from mass production and mass consumption to a new model that enables sustainable development. As a novel information technology, ubiquitous computing provides new service models and business opportunities for the manufacturing industry. This has given birth to the concept of ubiquitous manufacturing [13].

Ubiquitous manufacturing (UM), also known as ubiquitous factory, is an application of ubiquitous computing in the manufacturing sector. The name suggests that one could control a whole production and retailing chain from anywhere in the world and at any time. The key achievement of the application of ubiquitous computing in ubiquitous manufacturing is to provide the means for interconnected, smart objects. These smart objects can make the lives of users easier through automation. UM provides an environment in which manufacturing is done everywhere and anywhere.

The concept of ubiquitous manufacturing (UM) is a nontraditional classification of industry and the inclusion of market-oriented manufacturing. The concept carries the idea that one can control a whole production and retailing chain from anywhere in the world and at any time. The emergence of some advanced manufacturing innovations and technologies, such as lean manufacturing, cloud manufacturing, manufacturing grid, global manufacturing, virtual manufacturing, agile manufacturing, Internet manufacturing, sustainability manufacturing, smart manufacturing, and additive manufacturing has contributed to ubiquitous manufacturing [14]. UM does not imply that products could be supplied ubiquitously, since manufacturing a product ubiquitously is impossible.

Ubiquitous manufacturing is the next-generation manufacturing paradigm. UM enables ubiquitous, convenient, on-demand access to a shared pool of manufacturing resources. It is one of the emerging production paradigms toward which manufacturing companies are evolving. In recent years, ubiquitous manufacturing has attracted great attention in both academia and industry [15, 16].

Ubiquitous manufacturing must meet the following four conditions, known as the location controls of UM [2]:

1. Transporting finished goods is more expensive than transporting raw materials.
2. Finished goods are larger, more fragile, or more perishable than raw materials.
3. Raw materials are available ubiquitously.
4. Manufacturers and their customers communicate closely.

15.4 Related Technologies

Advances in computer science and engineering, such artificial intelligence, context-aware applications, ubiquitous computing, cyber-physical systems (CPS), Internet of Things (IoT), Internet of Services (IoS), big data, cloud computing, virtualization, and semantic networks are making their way into the industry. Unfortunately, these applications are isolated and not working together [17].

The emergence of advanced manufacturing technologies has led to ubiquitous manufacturing. Such technologies include lean manufacturing (LM), cloud manufacturing (CM), global manufacturing (GM), virtual manufacturing (VM), Internet manufacturing (IM), Industry 4.0, continuous manufacturing, agile manufacturing, additive manufacturing, chaordic manufacturing, etc. We hereby consider some of these technologies [18–24].

- *Cloud manufacturing* (CM): Recently, a cloud-based system is developed for ubiquitous manufacturing. Cloud manufacturing may be regarded as a ubiquitous manufacturing system. Cloud technology offers on-demand service access and resource pooling in the computing market. CM enables ubiquitous, convenient, on-demand network accesses to a shared pool of manufacturing resources, while UM emphasizes the mobility and dispersion of manufacturing resources and users. The cloud and ubiquitous manufacturing systems require effectiveness and permanent availability of resources. Both CM and UM grant factories unlimited production capacity and permanent manufacturing service availability. Cloud manufacturing represents such a recent transformation where multiple cloud models sustain the needed ubiquity of resources. Cloud provides new service models and business opportunities for the manufacturing industry.
- *Smart manufacturing:* There is a similarity between smart manufacturing systems and ubiquitous robotic systems, and effort is in progress to integrate ubiquitous robotic technology into the smart factory. Compared to conventional manufacturing processes, smart manufacturing offers the advantage of distributed networked machines to complete different tasks through collaboration. In contrast to the service robotic domain, smart manufacturing systems are often in larger size.
- *Industry 4.0:* Currently, we are experiencing the latest industrial revolution, known as Industry 4.0. Industry 4.0 may be regarded as the implementation of next-generation robotics for industrial applications. Ubiquitous manufacturing has been the most realizable implementation of Industry 4.0. The future manufacturing industry requires that the system could dynamically schedule the tasks for these machines according to their workloads and the received tasks [25]. Fig. 15.2 shows the components of Industry 4.0 [26].

Ubiquitous manufacturing can handle many problems that have not been properly solved by previous manufacturing paradigms. It suggests a nontraditional classification of industry and the inclusion of market-oriented manufacturers [27].

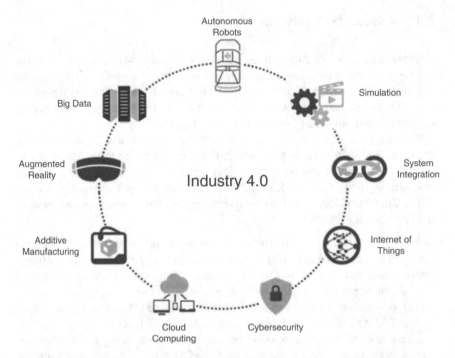

Fig. 15.2 Components of Industry 4.0 [26]

15.5 UM Characteristics

The application of ubiquitous technology is relatively new in the manufacturing sector. UM has the following characteristics [28]:

1. UM is a "design anywhere, make anywhere, sell anywhere, and at any time" concept that allows manufacturers an unlimited production capacity and permanent service availability.
2. It applies integrated manufacturing technology (MT), information technology (IT), and ubiquitous technology (UT) to the manufacturing domain. MT is for effective production, IT is for communication, while UT is for data acquisition and transmission. The ubiquitous environment in the manufacturing domain is characterized by being capable of producing product-centric digitized information.
3. Ubiquitous computing technologies address a wide range of issues in the manufacturing industry. Such issues include manufacturing processes and equipment, manufacturing management, and planning.
4. UM is a web-based manufacturing system, which can improve the efficiency and quality of product design, production, life cycle integration, enterprise manage-

ment, and customer service. It transparently collects and utilizes data on product and product-related context.

5. It supports real-time collaborative activities between stakeholders in a distributed environment.
6. UM may be regarded as wireless manufacturing or e-manufacturing.

15.6 Enabling Technologies

Advances in electronics (e.g., sensors, mobile devices, information, and communication technology) have significantly contributed to the emergence of ubiquitous manufacturing. UM uses technologies such as personal digital assistant (PDA), web camera, global position system (GPS), ZigBee, mobile Internet, USN (Ubiquitous Sensor Network), and autonomous industrial mobile robots. It employs wireless devices integrated into the products, manufacturing machines, and the factories themselves. A factory can be small or complex. Thus, UM enabling technologies include [29]:

1. *Sensors:* Ubiquitous sensing technologies include radio frequency identification (RFID), auto ID, virtual reality, GPS, and Wi-Fi. Sensors, machines, equipment, computers, products, etc., are equipped with embedded intelligence which makes them smart. A regular sensor can be converted into a ubiquitous sensor by connecting a networking module to it. The sensors (temperature, vibration, force, etc.) are used to monitor manufacturing processes. GPS has been used extensively in tracing the location of a delivered order. Wi-Fi has been applied for ubiquitous positioning, diagnoses, and control inside a factory.
2. *RFID*: RFID (radio frequency identification) is a widely adopted wireless sensor technology in many industries. Manufacturing resources like machines, materials, and personnel are equipped with RFID devices and they become smart manufacturing objects. RFID technology enables real-time traceability, visibility, and interoperability in improving the performance of shop-floor planning, execution, and control of manufacturing systems.
3. *Robots*: Industrial mobile robots are used in handling and transporting materials. This is appropriate for UM due to its flexibility and ability to communicate.
4. *Cloud Manufacturing*: This represents a shift from production-oriented to service-oriented manufacturing. It extends and adopts the concept of cloud computing for manufacturing. It is applied extensively to manufacturing because it enables collaborative product design by designers in different locations.

Other underlying technologies include operating systems, Internet, microprocessors, artificial intelligence, mobile devices and mobile networking, middleware and software architecture, wireless sensor networks, machine-machine communication and human-machine interface, and user interfaces.

15.7   Applications

Ubiquitous computing still holds lots of expectations and unopened doors and possibilities. As ubiquitous computing gains more footing, more applications will surface, some of which are applicable to ubiquitous manufacturing as well. Some common applications of ubiquitous manufacturing are presented as follows.

- *Automation:* Apart from object and process automation, enhanced human-machine interaction allows better safety, when, for example, machines stop when a human comes too near. Of course, automatic safety will not be totally foolproof, as machines might need some time for stopping, and stopping is not always possible. The use of ubiquitous technology in manufacturing leads to automation. Increase in automation on the manufacturing floor reduces human intervention and requires fewer workers.
- *Robotic Technology:* A robot is the prototype of an actuator and is controlled by the "intelligence" of an information system. Robots assemble, bolt, and fuse components, supervise oil drillings, disarm mines, or mow grass. The use of robotic devices, powered by ubiquitous technology, will enable a fully autonomous factory, with zero waste. An important feature of ubiquitous robotic systems is the development of a task-level learning and planning module that handles various tasks without recoding the robots. The ubiquitous robotic technology is mainly applied in the service robot domain. In a ubiquitous robotic system, robotic devices are often developed into modules that are connected through a network, enabling data sharing. A typical industrial manufacturing system can take advantage of the ubiquitous robotic technology. Some industrial robots are depicted in Fig. 15.3 [30].
- *Automotive and Aviation Industry:* The automotive industry is one of the leading sectors in the use of ubiquitous technologies. The automotive and aviation

Fig. 15.3 Some industrial robots [30]

industries use ubiquitous technologies to automate their processes and enhance production. The driving forces in the automotive and aviation industry include: increasing customer requirement, efficient design of the supply chain, product piracy, and recycling. Today, a car is more than a means of transportation for customers. Besides safety and comfort, customers expect some luxuries such as navigation, entertainment, and communication [31].

- *Smart Factory:* A factory is considered smart when all machinery and equipment can improve processes through automation and self-optimization. A smart factory can be implemented based on ubiquitous robotic technology. Unlike the traditional industrial robots that unintelligently or mindlessly repeat some predetermined tasks, robotic components in a smart factory are capable of sensing the environment and making decisions in the optimization of resources and time. However, the challenges of task planning for smart factory domains are introduced by their large problem size and uncertainty. As the manufacturing tasks are more individualized and flexible, machines in smart factories must perform variable tasks collaboratively. Figure 15.4 illustrates a typical smart factory [32].
- *Additive Manufacturing:* This is also known as 3D printing. It refers to building components by depositing material layer upon layer. The use of additive manufacturing will be ubiquitous in the aerospace and defense industry. Additive manufacturing will have a significant impact on the design of complex defense

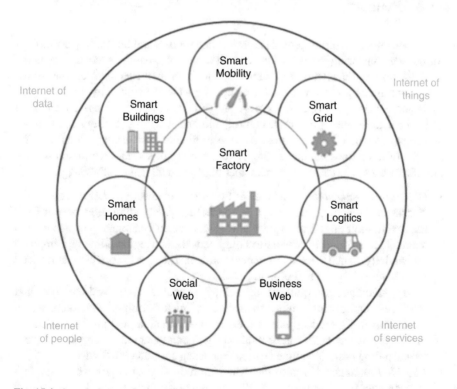

Fig. 15.4 A typical smart factory [31]

and aerospace products that break existing engineering limitations. Therefore, AM's penetration into the aerospace and defense value chain is poised to increase. The main advantage of additive manufacturing is its cost-effectiveness since there is no waste and the ability to design new components not before possible. It allows one to fabricate complex geometries that are not possible through traditional manufacturing techniques.

- *Manufacturing Executive Systems:* This is also known as collaborative production management. The basic concept of MES is to add functions to planning, logging, and control that act and react in real-time. Although MES are not directly related to ubiquitous manufacturing, they are enabled by the information generated by ubiquitous systems. The elements of an integrated MES are complete technical product description, resource management, planning and order management, performance monitoring, performance data recording, and information management. Therefore, MES can easily be seen as a central component in ubiquitous manufacturing systems [33].

Other areas of applications of ubiquitous manufacturing include the pharmaceutical industry, mobile communication, and the telecommunications industry.

15.8   Benefits

Ubiquitous manufacturing provides new service models and business opportunities for the manufacturing industry. The key advantage of ubiquitous technology is the digitalization and availability of more product data than ever before. The use of ubiquitous technology in manufacturing operations leads to ubiquity. Product quality is another competitive advantage. A big part of ubiquitous manufacturing is to make the manufacturing line more controllable. Introducing a UM service in convenience stores would help the store achieve profitable business opportunities. The benefits of ubiquitous manufacturing have taken the industrial automation revolution "Industry 4.0" to new horizons. Other benefits include the following [16]:

- *Ubiquitous Connectivity:* The idea of ubiquitous connectivity helps broaden the discussion to include other ways of connecting. Many companies benefit from improving their engagement with always-connected customers. For example, a manufacturer connects sensors to equipment to reduce downtime and improve productivity. Sensors are no longer exceptional. They are everywhere, embedded in all kinds of objects and even in people.
- *Cost Reduction:* Ubiquitous technology produces connected devices and machines. Optimizing supply-chain issues, manufacturing, and maintenance issues drastically reduces wastage, energy consumption, and production downtime. Areas where costs can be reduced significantly include energy cost, inventory handling cost, downtime cost, maintenance cost, and labor cost.
- *Flexible Production*: To meet the dynamic production and market demand, factory logistics must support the flexible production requirements. The ubiquitous

nature of ubiquitous manufacturing ensures to meet production flexibility. Ubiquitous technology is good at adapting to fit demand. Intelligent processes and tools enable faster, more cost-effective, quality-assured, and flexible production. This increases competitiveness in the global market environment.

- *Smart Factory:* Efforts have been made to deploy ubiquitous robotic technology to the smart factory. This grants factories an unlimited production capacity and permanent manufacturing service availability. The smart factory should adjust product type and production capacity in real-time.
- *On-Demand Access:* Ubiquitous manufacturing supports a demand-driven manufacturing approach. It provides an environment that enables on-demand network access to pooled configurable manufacturing resources and thus helps the platform of manufacturing everywhere and anywhere. Using UM, products can be supplied ubiquitously.
- *Quality Improvement:* Ubiquitous manufacturing improves the throughput of the factory and the quality of the product. Improvement in product quality often leads to lower wastage, higher customer satisfaction, and improved sales. Improved quality product implies zero defects or error in the manufacturing process.
- *Energy Efficiency:* Manufacturing operations consumes a lot of energy. Carefully monitoring energy consumption and the wastage in every production stage is critical to saving energy and cost. Use of ubiquitous technology is very effective in monitoring, tailoring, and optimizing the energy consumption of each machine.
- *Production Efficiency*: The data collected from ubiquitous manufacturing helps operators to visualize the production efficiency and wastages and foresee any arising issues. Ubiquity allows the manufacturer to have better insights into supply-chain issues, delivery status, and production efficiency matters.
- *Improved Productivity*: Today, the demand for manufacturing products is characterized by small batches with varieties. In contrast to traditional manufacturing, ubiquitous manufacturing can produce small-lot products of different types more efficiently. As the production process is optimized, the average manufacturing routes are shrunk, and the utilization rate of machines and other resources is improved.
- *Reduce Workforce:* The use of ubiquitous technology in manufacturing leads to automation, which reduces human intervention and requires fewer workers. Other workers can focus on other responsibilities.
- *Integration of Existing Systems*: The ubiquitous manufacturing system is not an island. It comes from the integration of existing systems and other emerging technologies such as cloud manufacturing and additive manufacturing.
- *Customization:* Ubiquitous manufacturing supports mass customization through interchangeable parts and configurations. This is enabled by having the information of customization travel with the product in the manufacturing phase. When information about the customization is carried in, it is also easy to track the custom orders.

These benefits are shown in Fig. 15.5.

Fig. 15.5 Some benefits of ubiquitous manufacturing

15.9   Challenges

Although ubiquitous manufacturing is a reality today, we are still wrestling with the challenges. As more devices are being equipped with wireless communication interfaces, the frequency bands allocated for this purpose will soon be exhausted. Many UM applications will only be successful when they are based on vendor-independent open standards [2]. One challenge is to determine which kind of real-time information should be provided for adaptive decision-making and how to deliver the information with existing information technologies [29]. The selection of appropriate cloud services for big data analytics is also a challenge [34]. Obviously, the longer it takes to manufacture a product, the more expensive it is. Other main challenges of ubiquitous manufacturing can be summarized as follows [25, 35]:

- *Security:* Today we are a data security-conscious society. Ubiquitous manufacturing is a model that is heavily information or data oriented. The privacy and the security of manufacturing data are also issues to be addressed. Manufacturers often put cost and time-to-market ahead of ensuring they comply with security best practices. Users, enterprises, and governments are increasingly concerned about privacy. Deployments that rely on ubiquitous connectivity can face significant challenges since unintended consequences of ubiquitous connectivity occur on a daily basis. Hackers present security problems as they target anything connected to the web. Ensuring security implies protecting individual system components against external attack and safeguarding the data transferred during communication procedures.

- *Heterogeneous Systems:* The robotic components of the ubiquitous robotic systems are highly heterogeneous with respect to platforms such as operating system, programming language, and communication media. People are faced with the reality of users that are increasingly mobile and need access to multiple, heterogeneous devices at different times and locations.
- *Complex Operations:* It is a big challenge to develop big data-based UM application systems to deal with the high volume, heterogeneous data produced by complex manufacturing operations. The manufacturing problems are often in large size with uncertainties.
- *Problems with RFID:* Challenges still exist in applying RFID technologies to real-life UM environment. It is both labor and skill intensive to integrate and manage various kinds of RFID devices for various industrial applications. Requirement changes of applications is a challenge for RFID system in the manufacturing environment [36].
- *Lack of Awareness:* Manufacturers are unaware of how ubiquitous manufacturing can benefit them. As a result of this lack of awareness in the industrial community, most manufacturers are reluctant to adopt ubiquitous manufacturing.
- *Expensive Maintenance:* The continuous maintenance of ubiquitous facility (leading to automation and virtualization) is expensive. Ubiquitous devices installed in several places are delicate and subject to damage due to industrial heat, moisture, dust, physical impacts, etc. Their damage can cause business downtime, which makes adopting ubiquitous manufacturing difficult for manufacturers to accept.
- *Lack of Standard:* The absence of industry standards is another major challenge of UM. In the manufacturing environment, la ack of standardization represents a big risk. This is why standards (e.g., for data exchange protocols) are necessary. Ubiquitous technologies are embedded into everyday things that vary greatly from each other. This means that in terms of standardization, implementing ubiquitous computing concepts into manufacturing should be much easier than into customer goods.

These challenges are portrayed in Fig. 15.6. They are open for investigation to improve work-cell productivity and quality in a UM environment facilitated with RFID technology.

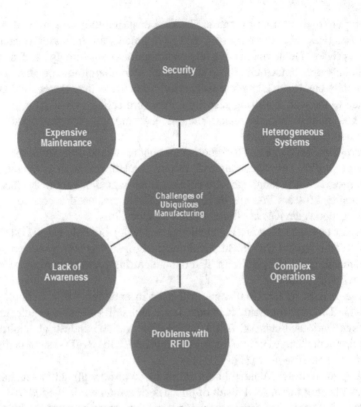

Fig. 15.6 .Some challenges of ubiquitous manufacturing

15.10   Global Ubiquitous Manufacturing

The name "ubiquitous manufacture" carries the idea that one could control a whole manufacturing and retailing chain from anywhere in the world and at any time. The objective is to implement technically a new way to do business. The technology developed during the project addresses several of globalized companies. Several software solutions exist, but they still bare crucial limitations and do not support manufacturing processes in a broader sense. We now consider how ubiquitous manufacturing is applied in some nations.

- *United States:* A US company Baker Tilly has implemented services to connect and automate manufacturers' production. NEST (which is now owned by Google) includes everything from automatic energy and use regulation when on holiday to handy temperature scheduling systems. It is easy to see why the popularity of ubiquitous devices is on the rise. Intel designed a more geographically diversified, sustainable, and secure semiconductor supply chain. This investment will create jobs, foster innovation projects, and reposition the United States as a technology and manufacturing leader [37].

- *Korea:* ACS (Advanced Consulting Services), a Korean company based in Seoul's digital city, was the first high-tech complex in the world for digital technologies, which put theories into practice. In 1988, ACS was the first to implement a manufacturing software designed for global operation in South Korea. Since then, Korea became the innovation hub. The Korean touch Cha likes to demonstrate on a board having pride of place in his office the functioning of the modern global company: "with our technology you can easily coordinate design in Korea, production in Brazil, planning in the USA, maintenance in France and recycling in China for a product actually distributed and sold in Germany" [17].
- *Japan:* Fujitsu Limited is the leading Japanese information and communication technology (ICT) company, offering a full range of technology products, solutions, and services. It has been pushing ahead with company-wide initiatives on many fronts, from the cloud, platforms, and networks to front-end devices. In 2015, Fujitsu announced that it would strengthen its Ubiquitous Solutions business. The objective is to develop competitively, produce superior products, and expand its business. This will enable the company to provide the market with even more competitive and timely new products [38].
- *Europe:* The manufacturing sector in Europe must respond to demands for low and variable volumes of high quality, at low cost (in terms of time, energy, space, and price). This is of particular relevance to the aerospace sector. Such production systems must deal with the challenge of increasing demand and regulation. The key challenge is then the transformation of manufacturing systems into dynamic facilities that can self-learn, self-adapt, and self-reconfigure [39].

15.11 Future of Ubiquitous Manufacturing

Today's manufacturing will surely not be the manufacturing of tomorrow, as it may be too rigid to adjust to the demands of a globalized market. In the factory of the future, all the pieces, tools, objects, and people know where to be and when.

There are several manufacturing paradigms competing for the future, and it is too early to predict which one will succeed. Ubiquitous manufacturing is one of the emerging production models toward which manufacturing companies are evolving. Ubiquitous manufacturing technologies have been widely deployed in a variety of manufacturing processes. The trend is to integrate logistics planning with the UM task scheduling using big data analytics of the manufacturing data. We are gradually evolving from hierarchical to decentralized decision-making. This evolution in turn will provide new opportunities.

15.12   Conclusion

Manufacturing activities have evolved over the years. With the convergence of wireless technologies, advanced electronics, and computer networks such as the Internet, the scope of manufacturing has expanded beyond geographical boundaries. Ubiquitous manufacturing basically means the usage of ubiquitous computing concepts and technologies in manufacturing. It implies product design, manufacturing, and recycling via ubiquitous computing technology. It introduces changes to the factors and elements of traditional manufacturing systems and incorporates the current requirements of smart systems.

Ubiquitous manufacturing is a manufacturing paradigm which is characterized by "design, makes, and sell product anywhere and at any time." Since UM involves the whole lifecycle of a product, it should involve the collaboration of government, R&D centers, and enterprises. Government handles legislation, while R&D centers conduct research and development, and the enterprises invest in the technologies. More information about UM can be found in the book in [40].

References

1. J.S. Yoona, S.J. Shin, S.H. Suh, A conceptual framework for the ubiquitous factory. Int J Prod Res **50**(8), 2174–2189 (2012)
2. T. Chen, H.R. Tsai, Ubiquitous manufacturing: current practices, challenges, and opportunities. Robot Comput Integr Manuf **45**(C), 126–132 (June 2017)
3. M.N.O. Sadiku, Y. Wang, S. Cui, S.M. Musa, Ubiquitous Computing: A primer. Int J Adv Scientif Res Eng **4**(2), 28–31 (2018)
4. S. Singh, S. Puradkar, Y. Lee, Ubiquitous computing: connecting pervasive computing through semantic web. IseB **4**(1), 421–439 (2006)
5. M.N.O. Sadiku, S.M. Musa, O.S. Musa, Pervasive computing. Invent J Res Technol Eng Manage **1**(12), 30–32 (2017)
6. G.D. Abowd, *Beyond Weiser: from ubiquitous to collective computing*, vol 49 (Computer, 2016), pp. 17–23
7. M. Weiser, The computer for the twenty-first century. Sci Am **265**(3), 94–100 (1991)
8. M. Weiser, Some computer science issues in ubiquitous computing. Commun ACM **36**(7), 75–84 (1993)
9. http://www.ubiq.com
10. Features of cloud computing – 10 major characteristics of cloud computing, https://data-flair.training/blogs/features-of-cloud-computing/
11. Ubiquitous computing, https://www.techopedia.com/definition/22702/ubiquitous-computing
12. J.S. Winter, Emerging policy problems related to ubiquitous computing: negotiating stakeholders' visions of the future. Knowl Technol Policy **21**(4), 191–203 (2008)
13. Ubiquitous Manufacturing (UbiM), Robotics and Computer-Integrated Manufacturing (2015)
14. S.H. Suh, UbiDM: A new paradigm for product design and manufacturing via ubiquitous computing technology. Int J Comp Integ Manuf **21**(5), 540–549 (2008)
15. M.N.O. Sadiku, E. Dada, S.M. Musa, K.B. Olarenwaju, Ubiquitous Manufacturing. J Scient Eng Res **7**(5), 212–214 (2020)
16. M.N.O. Sadiku, O.D. Olaleye, A. Ajayi-Majebi, S.M. Musa, Ubiquitous Manufacturing. Int J Trend Res Develop **8**(1), 10–11 (2021)

17. The ubiquitous factory, https://cordis.europa.eu/news/rcn/129833/en
18. D. Zuehlke, *SmartFactory* – A vision becomes reality, *Proceedings of the 13th IFAC Symposium on Information Control Problems in Manufacturing*, Moscow, Russia, June 2009
19. R. Dubey, A. Gunasekaran, A. Chakrabarty, Ubiquitous manufacturing: overview, framework and further research directions. Int J Comp Integ Manufact **30**(4–5), 381–394 (2017)
20. W. Wang et al., Ubiquitous robotic technology for smart manufacturing system. Comput Intell Neurosci **2016**, 6018686 (2016)
21. P.K.D. Pramanik et al., Ubiquitous manufacturing in the age of industry 4.0: A state-of-the-art primer, in *A Roadmap to Industry 4.0: Smart Production, Sharp Business and Sustainable Development*, ed. by A. Nayya, A. Kumar, (Springer, 2020), pp. 73–112
22. Additive manufacturing will be 'ubiquitous' in defence within a decade, survey indicates, https://www.defenceiq.com/defence-technology/articles/additive-manufacturing-will-be-ubiquitous-in-defen
23. X.V. Wang et al., *Ubiquitous manufacturing system based on cloud: a robotics application* (Robot Comp Integ Manufact, 2016)
24. M. Cheng et al., Cloud service-oriented dashboard for work cell management in RFID-enabled ubiquitous manufacturing, *Proceedings of the 10th IEEE International Conference on Networking, Sensing and Control*, 2013, pp. 379–382
25. P.K.D. Pramanik et al., Ubiquitous manufacturing in the age of Industry 4.0: A State-of-the-Art Primer, in *A Roadmap to Industry 4.0: smart production, sharp business and sustainable development. advances in science, technology & innovation*, ed. by A. Nayyar, A. Kumar, (Springer, 2020)
26. T. Melanson, What Industry 4.0 means for manufacturers, https://aethon.com/mobile-robots-and-industry4-0/
27. Y.C. Lin, T. Chen, A ubiquitous manufacturing network system. Robot Comput Integr Manuf **45**, 157–167 (2017)
28. J.B. Foust, Ubiquitous manufacturing. Ann Assoc Am Geogr **65**(1), 13–17 (March 1975)
29. I. Nielsen, Material supply scheduling in a ubiquitous manufacturing system. Robot Comput Integr Manuf **45**, 21–33 (2017)
30. Industrial robotics: An introduction and beginner's guide, June 2019, https://www.rg-robotics.com/industrial-robotics-an-introduction-and-beginners-guide/
31. E. Fleisch, Ubiquitous network societies: their impact on the telecommunication industry, 2005., https://www.researchgate.net/publication/227984950_Ubiquitous_Network_Societies_Their_Impact_on_the_Telecommunication_Industry
32. What are the benefits of a smart factory? https://www.quora.com/What-are-the-benefits-of-a-smart-factory
33. J. Kiirikki, Ubiquitous manufacturing requirements for product data management, *Masters Thesis,* Tampere University of Technology, (December 2010)
34. X. Wang, S.K. Ong, A.Y.C. Nee, A comprehensive survey of ubiquitous manufacturing research. Int J Prod Res **56**, 1–5 (2017)
35. Forget the IoT – Digital business needs ubiquitous connectivity, https://isg-one.com/research/articles/forget-the-iot-digital-business-needs-ubiquitous-connectivity
36. J. Fang et al., Agent-based gateway operating system for RFID-enabled ubiquitous manufacturing enterprise. Robot Comput Integr Manuf **29**, 222–231 (2013)
37. The importance of investing in domestic manufacturing, https://www.intel.com/content/www/us/en/corporate/usa-chipmaking/home.html
38. Fujitsu strengthens its ubiquitous solutions business, https://www.fujitsu.com/global/about/resources/news/press-releases/2015/1029-02.html
39. D. Sanderson, J.C. Chaplin, S. Ratchev, Conceptual framework for ubiquitous cyber-physical assembly systems in airframe assembly? *IFAC-PapersOnLine* **51**(11), 417–422 (2018)
40. C.T. Toly, *3D printing and ubiquitous manufacturing* (Springer, 2020)

Chapter 16
Offshore Manufacturing

Do what you do best and outsource the rest.

–Peter Drucker

16.1 Introduction

Manufacturing refers to any business practice that transforms raw materials into finished or semifinished goods. The manufacturing industry is the major contributor to any nation's economy. The globalization of manufacturing is well established in many industries. It is changing the nature of international trade. The global ever-increasing competition and fast technological change lead companies to combine their resources with those available in other countries. However, there has been a negative reaction to globalization based on the US manufacturing employment decline. Many multinational enterprises in the United States experienced job losses. Increasing global competition and profit-maximizing objectives has led many firms to expand to offshore jurisdictions. Offshoring (or offshore outsourcing) refers to relocating any business process abroad and then selling the goods in the company's home country.

Offshoring manufacturing takes place when production operations are performed in another country, including manufacturing and product assembly, from one country to a low-cost country. It is considered an option when a company wants to reduce operating expenses while maintaining high-quality and sometimes compromised-quality products. The concept of offshoring is a business practice that has gained momentum and has become a widespread business practice among companies worldwide [1].

The counterintuitive and definitely questionable major business principle behind offshoring is that it is not better to be sending goods 2 km away than shipping them 2,000 km across political boundaries in 24 days. Keeping plants near customers shortens lead time. Experience shows that many offshore manufacturers tend to overestimate profit and overlook problems such as currency exchange rate and political instability of the offshore locations [2].

© The Author(s), under exclusive license to Springer Nature Switzerland AG 2023
M. N. O. Sadiku et al., *Emerging Technologies in Manufacturing*,
https://doi.org/10.1007/978-3-031-23156-8_16

This chapter provides an introduction to offshore manufacturing. It begins by explaining what offshore manufacturing is all about. It compares local/domestic manufacturing with offshore/foreign manufacturing. It also compares offshoring and outsourcing. It addresses offshore locations such as China, Mexico, and India. It highlights the benefits and challenges of offshore manufacturing. The last section concludes with comments.

16.2   What Is Offshore Manufacturing?

Offshore manufacturing (OM) (also known as offshore outsourcing or production offshoring) involves the relocation of manufacturing facilities and processes overseas in a country where labor and/or raw materials are cheaper. Offshore manufacturing is part of the umbrella term "offshoring," which means the relocation abroad of any business process or operation in a foreign country, usually with the strategic intent of a tax haven. Conventional wisdom has it that companies go offshore to reduce costs, usually to low-wage countries. Factors that dictate offshoring can be technological, economic, geopolitical, cultural, psychosocial, regulatory, strategic, or an array of other influencing factors [3].

Although offshoring, opportunities for developed and developing countries, is an age-old phenomenon, many Western multinational establishments started offshoring part of their activities and procurement in the 1980s. Today, there are literally thousands of companies of all sizes that outsource part or all of their manufacturing to overseas companies. Popular companies that opt for offshore manufacturing include Apple, Nike, Cisco, Walmart, and IBM [4]. An example of offshoring manufacturing is illustrated in Fig. 16.1 [5]. As another example, offshore oil production

Fig. 16.1 An example of offshoring manufacturing [5]

Fig. 16.2 Offshore semisubmersible oil production platform [6]

platforms include production separators for separating the produced oil, water, and gas, as illustrated in Fig. 16.2 [6].

There are four basic types of offshore outsourcing [7]: (1) information technology outsourcing, (2) business process outsourcing, (3) offshore software development, and (4) knowledge process outsourcing. Services such as call centers, computer programming, reading medical data such as X-rays and magnetic resonance imaging, medical transcription, income tax preparation, title searching, and software, electronic, and pharmaceutical intellectual property are being offshored.

Offshore manufacturing is essentially relocating the production of goods to another country. It means relocating manufacturing abroad and then selling the goods in the company's home country. The driving factors behind offshoring include the following [8, 9]:

- *Cost Savings:* The primary motivating factor for companies to move their production offshore is to save money. Offshoring is regarded as an opportunity to reduce total transaction costs and production costs. The ultimate goal is to help your customers remain extremely competitive. Your competitors may not have facilities overseas and thus may have higher costs.
- *Competitive Concerns:* To be competitive on the global stage, corporations need globally competitive prices. Offshore manufacturing can allow companies to remain competitive in an ever-changing market. Organization flexibility in manufacturing is increasingly critical for companies to remain competitive in the marketplace. Offshore enhances international competitiveness so as to reduce costs, expand globally, serve customers more effectively, free up scarce resources, and leverage capabilities of foreign partners [10].
- *High-Volume Manufacturing*: Production volume is key. Offshore production facilities are often designed to support high-volume production. This can be

attributed to the fact that overseas manufacturing is less expensive than domestic production.

- *Lower Labor Costs:* Cheap labor is almost synonymous to offshore manufacturing. Offshoring allows you to reduce labor cost, which is one of the most expensive parts of your business. Overseas labor is only a fraction of the cost of domestic facilities labor operational costs. Businesses with more labor-intensive products are bound to see the biggest return on investment. Although low labor costs are often the primary reason for companies to look abroad, a company must be able to think beyond labor costs.

- *Contract Manufacturing:* A contract manufacturer is a manufacturer that contracts with a company for products. It is one form of outsourcing. The contract manufacturer has full responsibility for component manufacturing or complete assembly of finished products. In this type of configuration, the company establishes the requirements that the contract manufacturer must meet in relation to quality of the products, packaging specifications, the lead time, and the delivery schedule. Today, contract manufacturing is embraced by multinationals like IBM, Philips, Lucent, Texas Instrument, and Ericson. Questions to ask in determining the value a contract manufacturer can deliver in this area include [11]:

  - What regulatory bodies does the contract manufacturer work with most frequently?
  - What resources does the contract manufacturer have dedicated to quality management and compliance activities?
  - Is there a well-defined process for transfer of work?
  - How does the contract manufacturer address tooling compatibility issues?
  - Can the contract manufacturer's engineering team support product redesign for cost reduction or added functionality?
  - Does the contract manufacturer have the support staff and processes in place to support product qualification and production validation processes?
  - Does the contract manufacturer's team look at business issues such optimum logistics strategy?

These terms are also used to describe offshoring manufacturing.

Various industries including automotive, computer, food manufacturing, aerospace, packaging, financial services, telecommunications, and others utilize offshore manufacturing to increase production while lowering costs. For example, BMW, a German car manufacturer, may relocate a BMW factory in the United States in order to easily reach the American auto industry. British pharmaceutical company AstraZeneca (AZ) has production facilities in China and produces goods for Chinese consumers. American carmaker Ford has factories in Mexico with Americans consuming most of the cars. Japanese multinational enterprises have steadily increased their offshore manufacturing presence.

16.3   Local or Offshore Manufacturing

The choice between domestic/local and offshore/foreign manufacturing is sometimes not easy. In order to make a clear and informed decision, it is important to know the pros and cons for both. While local manufacturing is more accommodating to smaller quantity, offshore manufacturing is generally about large quantity production. Although the price from an offshore manufacturing organization is lower than its local counterpart, it is important to factor in some high costs such as transportation costs [12].

When designing a new product, a business must decide whether to work with a domestic manufacturer or outsource the production overseas. It is not always true that local manufacturing is more expensive than outsourcing production overseas. The perception that offshore manufacturing is cheaper may not take into account hidden costs such as shipping, tariffs, and taxes. For inventors that want to manufacture products on a small scale, there may be no cost benefits when they outsource this process offshore. For example, demand for computers, cell phones, accessories, and the related devices make them prime candidates for nearby sourcing. Local manufacturers have the following advantages [13]:

- Localizing supply chains
- Greater supply chain stability
- High-quality assurance
- Less waste
- Fair working conditions
- Adherence to strict safety standards
- Keeps intellectual property closer to home
- Reduced shipping costs
- The "Made in the USA" label does sell products
- Makes communication much easier
- Offers greater flexibility

If you are a manufacturer and local manufacturing is expensive, to remain competitive, then you may consider offshore manufacturing.

Some companies outsource certain components, parts, and processes that are not among their core competencies. They outsource as a strategy to strengthen their competitive advantage. Other specific reasons companies offshore their manufacturing include [14]:

- *Shortage of Skills:* A shortage of employees with the required skills is one of the main reasons why they offshore some of their manufacturing processes. Offshoring maximizes the efficiency of operations and boosts business productivity.
- *Cost-savings:* Factories offshore their production overseas to access lower manufacturing costs. It is estimated that companies can save up to 80% in cost savings.

- *Efficient Quality and Process:* Offshoring destinations in Asia are efficient in setting up systems that conform to standard rules and procedures.
- *Time zone coverage*: Companies can cover time zones not handled by their operations through subsidiaries or offshore service providers in countries offering 24/7 operations.

One must understand that offshore manufacturing is not all roses and labor savings. It may be faster, convenient, and cheaper to do business with a company that is just around the corner than one located halfway around the globe. It may be wise to outsource in the United States. Keep in mind that other cultures have different customs, but people are friendly just like everywhere else. Some cogent reasons not to consider offshore manufacturing include [15, 16]:

- *Not the Only Option:* Many American manufacturers claim that going offshore is their only alternative if they are to survive and stay competitive in the global market. They insist it is either offshore manufacturing or no manufacturing at all. However, this blatant argument may not seem to go far in face of several disadvantages or challenge facing offshore manufacturing. For example, more Japanese automakers and electronics companies are building manufacturing plants in the United States when many American companies are claiming it is impossible for them to stay home and be competitive. Japanese factories are competitive and successful because of their excellent manufacturing techniques management philosophy.
- *Loss of Intellectual Property*: Intellectual property (IP) is always an issue because businesses which offshore are bound to expose some of their IP to suppliers. Making the shift to offshore production might leave critical company secrets vulnerable and lead to the theft of intellectual property. IP theft has become an issue due to the exposure. US companies that transfer technology across national borders do not have the same protection against piracy they enjoy at home. IP theft has become an issue. China is well-known for its poor enforcement of intellectual property laws, and there is a serious risk of copycat products. Korea denies copyright protection to software, semiconductors, or foreign works.
- *Shipping:* Shipping is a major additional cost of offshore manufacturing. Manufacturing overseas may need to ship the finished products back home. This often involves finding reliable partners for shipping. Figure 16.3 shows an off-

		Location	
		Domestic	**Abroad**
Ownership	**In-House**	Domestic internal production	Vertical FDI / Production by foreign affiliate
	External	Domestic outsourcing	International outsourcing / arm's-length contracts

→ Outsourcing

↓
Offshoring

Fig. 16.3 Definition of offshoring and outsourcing [6]

shore container to transport equipment and supplies [17]. The ever-increasing costs of transporting (by ocean and by air) materials to and finished products from international locations may be high. The shipping cost reduces the advantage of offshore manufacturing.

- *Do Not Really Save Much:* Because labor is cheap, the Chinese have little incentive to improve process efficiency. Consequently, lean manufacturing is a more difficult process to implement in China than in the United States. There is the possibility of suffering great loss due to hidden costs in such areas as are impacted by cultural and political differences, warranties, and privacy issues.
- *Collaborators May Become Competitors*: Distance may make collaboration very challenging. The danger that collaboration will give way to competition is real.
- *Foreign Exchange*: Manufacturing abroad usually involves dealing with a different, foreign currency which may be fluctuating and may not favor the home business. Currency volatility and wage-rate inflation occur in the low wage country. This implies that the low wage jurisdiction will not be low wage forever.

16.4   Offshoring and Outsourcing

The rapid growth of offshoring and outsourcing has transformed the way businesses have managed their operations globally. There is often some misconceptions about the difference between offshoring and outsourcing. The two terms have been used interchangeably because they share some characteristics, yet they refer to different firm strategies and they can be combined. They are interrelated. Other related terms include onshoring, reshoring, inshoring, insourcing, backshoring, bestshoring or rightshoring, and nearshoring [7]. There are several factors that determine the extent of outsourcing and offshoring.

Outsourcing is an "umbrella" term. Offshoring takes place when you relocate the work to a different country and is a form of outsourcing. Offshoring moves production operations to another country, but the company remains in complete control of all aspects of the operation. Offshoring is always outsourcing, but not all outsourcing is offshore.

Figure 16.3 shows the definition of offshoring and outsourcing [6].

The term "outsourcing" means contracting out a portion of the business activity to an external party rather than doing it in-house [18]. Outsourcing manufacturing lets you scale up production to meet demand without a major investment. Lower cost and increased corporate profitability are often the major motivations. Things like customer satisfaction, product education, delays in time to market, overseas communication, the target country's level of economic development and political stability, project management, adjustment cost, transportation and logistics, cultural sensitivities, and language differences are just a few of the long-term hidden costs for outsourcing.

It is expedient for a manufacturer to outsource part of the manufacturing process to third parties, especially when it does not have the equipment required to fabricate certain components. Outsourcing is a prominent feature of many recent models of international trade. Outsourcing is an effective means by which a company can reduce overhead costs and increase profits. Outsourcing allows companies to mainly focus on sales and marketing, while entrusting a firm to handle production, quality control, and administration. Outsourcing increases efficiency and allows a company to accomplish the same task for less money [19]. For example, some companies outsource various professional services such as emailing, payroll, and call center.

A simpler and more effective outsourcing alternative is onshoring. Onshoring offers improved communication and increased productivity between both parties. For example, a company located in New York City can contract services to a company located in smaller city like Huntsville, Alabama, where living expenses are much lower.

In 1986, United Technologies closed its diesel-engine parts plant in Springfield, Massachusetts, and transferred operations to a plant in South Carolina and two plants in Europe. By working with a company located in the same country, both parties will benefit from more convenient time zones, faster and cheaper business travel, and easier collaboration. Onshoring eliminates the need to look overseas for quality software development outsourcing and keeps jobs in the United States. It also eliminates the problems of data security, political instability, time zones, distance, language barriers, etc. [20].

16.5   Offshore Locations

When considering offshore manufacturing, a first step is to determine the best country for the location of the offshore manufacturing. Common locations for offshoring include China, Vietnam, India, Brazil, Turkey, Malaysia, Taiwan, the Philippines, Pakistan, and other countries. US-based companies often consider China and Mexico as the widely utilized locations for offshoring manufacturing operations. China is the largest offshoring destination for US-based companies. Mexico may be a better option due to the close proximity of Mexico to the United States. Duties may be considered depending upon the governing tariff code. Shipping time, age, and availability of the labor force are important factors to think about [7]. Countries like China and India have become major sources of some services due to perceived lower costs stemming from cheap labor and production. Due to space limitation, we will consider three main countries where US companies make offshoring [21]:

- *China:* China is the largest exporter to the United States. It offers a potentially massive domestic market, second the United States. China's major advantage as a nation is low-cost labor. In China, the average hourly wage in China is significantly lower than Mexico. China had fewer workers' rights than Mexico, fewer environmental regulations, and poorer enforcement of intellectual property laws. Compared to offshore operations in China, the main advantages of operations in Mexico are shown in Fig. 16.4 [22]. Today China produces various items such as

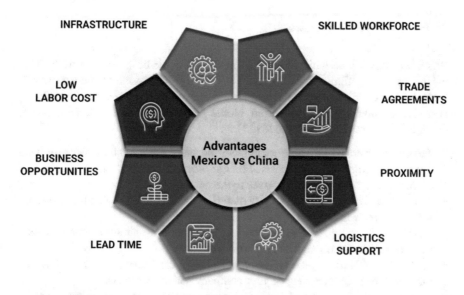

Fig. 16.4 Compared to offshore operations in China, the main advantages of operations in Mexico are shown [22]

clothing, equipment, electronics, and toys. China has work ethic but lack proficiency of English communication.

- *Mexico*: Mexico has a privileged geographical position with cheap labor. Besides labor being the primary cost savings in Mexico, the demographics of the country make it an ideal choice for US-based companies. The introduction of the North American Free Trade Agreement (NAFTA) has made it easier for American manufacturers to relocate to Mexico. A *maquiladora* was an assembly plant operated in Mexico under a preferential tariff program. It originated as part of the Mexican government's 1965 Border Industrialization Program, which targeted to stimulate the manufacturing sectors of depressed economies.

- *India:* India is a leader in relocation services and administrative processes. It has the advantages of English proficiency and a large number of skilled workforces especially in computer engineering. India's education system is mainly focused on higher education. State and central governments in India have offered companies financial incentives to attract foreign investors. Today, India is the prime location for offshoring IT operations such as code maintenance, help desk, and desktop maintenance. India's software outsourcers have attracted billions of dollars of US foreign direct investment. India is considered number one for developing drugs that are approved by the US Food and Drug Administration. The main challenge a company may face in India is cultural differences.

Besides the locations for offshoring manufacturing operations, important factors a company must consider include the wage rate of workers and whether the workforce can maintain the quality of the product.

16.6   Benefits of Offshoring

When it comes to international investment, offshoring is becoming the trend. This business practice offers a reduction of costs for the company and an increased quality in products. Offshore manufacturing can be a real boon. It provides convenience, cost savings, high volume production, global presence, and international markets. Other benefits of offshoring include [23]:

- *Lower labor costs:* Offshore manufacturing labor costs have historically been lower than domestic labor costs. Foreign workers benefit from new jobs and higher wages when the work moves to the country.
- *Less expensive raw materials*: Raw materials often cost less when the manufacturer is close to their source.
- *Easier access to specialized suppliers:* Manufacturing countries such as Asia often have a larger number of specialized suppliers.
- *More time to focus on product development:* You can focus on product development and the manufacturing of your company's niche products.
- *Eliminates operational concerns:* OM represents a net gain and reduces operational headaches.
- *Economic development:* OM is one of the main reasons for the economic development of the developing countries like China. It leads to economic growth of many developing countries, which has led to currency.
- *Local Representatives:* Having local representatives available in China, India, Malaysia, Vietnam, Indonesia, or Thailand to advocate for your company is critical in maintaining product quality, effective communication, and surveillance of offshore manufacturing site.

These benefits are illustrated in Fig. 16.5. Other benefits of manufacturing offshore include lower production, lower capital costs, and greater manufacturing capacity. In most cases, the benefits outweigh the risks. Notwithstanding these potential benefits, offshore manufacturing also brings potential disadvantages.

16.7   Challenges of Offshoring

Offshore manufacturing could be a profitable experience. However, some companies experience huge challenges, frustrations, and disappointments that lead to financial loss. When sourcing offshore manufacturing, stresses about quality deterioration, unreliable delivery, world economy, fluctuating exchange rates, political instability, and untrustworthy suppliers can be major issues to worry about. The rapid growth of offshoring has sparked a contentious debate over its impact on the US manufacturing industry partly due to the steep employment declines and numerous plant closures. Most companies made the decision to offshore based on miscalculations and high expectations. Offshoring often demands much more time, more management oversight, and a more nuanced approach than most companies realize. Other challenges of offshoring include [20, 24, 25]:

Fig. 16.5   Some benefits of offshore manufacturing

- *Risks Involved:* There are risks associated with offshore manufacturing. OM incur risks that have to with quality issues, shipping costs, duties, currency revaluations, and delays in time to market. A wide range of factors exist that may potentially affect quality and risk in offshore manufacturing. For example, a company can lose time, money, and reputation as it engages the practice of offshore manufacturing.
- *Quality Suffers*: Quality becomes an issue comparing onshore/nearshore and offshore manufacturing. Quality concerns can cause some rethinking about what goes offshore and what stays home. There is a wide range of factors that can affect quality in OM. Overseas suppliers may use inferior components.
- *Lack of Support:* US and state governments do not offer a lot of support for offshore manufacturing.
- *Job Loss:* OM is often blamed for job loss. Some believe that OM moves jobs to another country, thereby increasing unemployment in the home country. Ordinary citizens may be adversely affected by unregulated, offshore activities. Jobs are added in the destination country and are subtracted from the higher-cost labor country. In other words, there is job loss in the origin/home country and job gain in the destination country.

- *Time Zone Difference:* A major challenge of offshoring is time zone differences. Many offshoring companies operate within a 5–12 hour difference from their client.
- *Language Barrier:* When working with someone in another country who natively speaks another language, this can make communication and collaboration a major challenge. Language barriers and cultural differences add to the complexity.
- *Intellectual Property Concerns:* The security of your intellectual property can be a major concern. Protecting your intellectual property (IP) may be important than reducing your manufacturing costs.
- *Data Security:* This is an important factor for offshoring. It entails giving access to some classified customer data and information to offshore manufacturers. Any breach in data security could heavily impact your business and customers.

Other challenges include cultural and social differences, geopolitical uncertainty, and counterfeiters. As manufacturing costs in low-cost countries have increased, it becomes questionable whether it makes sense to offshore production. There are still many good reasons for US companies to continue to offshore manufacturing in Asia.

16.8   Conclusion

Globalization has made offshoring a predominant business practice in many large companies, and offshoring has become an irreversible trend. Offshoring is the movement of a business process from one country to a lower-cost country or one with fewer regulatory restrictions. It allows companies today to collaborate simultaneously with multiple partners across the globe. Manufacturers in the United States and Europe can make sense of offshore manufacturing. Offshoring can increase a company's domestic employment. When deciding whether to build new plants here or offshoring, offshoring may present a higher or a lower risk depending on a variety of important factors tied to strategic business decisions. Businesses benefit from accessibility to labor resources across the world. The types of products best suited for offshoring are products with high labor content. The common products that are manufactured offshore include car parts, electronics, furniture, petroleum, etc. Although it is still viable for US companies to source and manufacture in Asia, more careful cost-benefit analysis is required to arrive at a robust and wise business decision that delivers sustained business advantage in the long term.

To prepare the workforce, colleges need to adapt and provide graduates who have some educational background in offshore manufacturing. Some engineering programs are realizing the importance of international experience and incorporating offshore manufacturing in their programs [26]. Offshore manufacturing is here to stay, but it is not for everyone [27]. More information about offshore manufacturing can be found in the books in [28–34] and related journal: *Strategic Outsourcing: An International Journal.*

References

1. M.N.O. Sadiku, O.D. Olaleye, S.M. Musa, Offshore manufacturing: A primer. *Int J Trend Res Develop* **6**(3) (2019)
2. R.C. Ritter, R.A. Sternels, When offshore manufacturing doesn't make sense. *McKinsey Q* **4**, 124–127 (2004)
3. W.W. Chang, The economics of offshoring. *Global J Econom* **1**(2), 1250009 (2012)
4. Which five companies do the most overseas manufacturing? https://www.itimanufacturing.com/five-companies-overseas-manufacturing/
5. Offshore manufacturing – definition and meaning, https://marketbusinessnews.com/financial-glossary/offshore-manufacturing/
6. N. Bottini, Offshoring and the labour market: What are the issues? http://www.ilo.org/public/english/employment/download/elm/elm07-11.pdf
7. Offshoring, *Wikipedia,* the free encyclopedia. https://en.wikipedia.org/wiki/Offshoring
8. What are the benefits of offshore production? May 2020, https://www.casesbysource.com/blog/benefits-offshore-production
9. A. Mykhaylenko et al., Accessing offshoring advantages: What and how to offshore. *Strat Outsourc Int J* **8**(2/3), 262–283 (2015)
10. D.D. Gregorio, M. Musteen, D.E. Thomas, Offshore outsourcing as a source of international competitiveness for SMEs. *J Int Bus Stud* **40**, 969–988 (2009)
11. Offshore manufacturing: Risky business or the right choice for your project? https://forefront-medical.com/zh/offshore-manufacturing-risky-business-right-choice-project/
12. Offshore v's onshore production- the pros and cons, https://sampleroom.wordpress.com/2013/06/17/offshore-vs-onshore-production-the-pros-and-cons/
13. 5 Benefits of local manufacturing in the Pacific Northwest, https://www.pacific-research.com/5-benefits-of-local-manufacturing-in-the-pacific-northwest-prl/
14. Why do companies need offshoring, https://fullscale.io/why-do-companies-need-offshoring/
15. C. C. Markides and N. Berg, Manufacturing offshore is bad business, September 1988, https://hbr.org/1988/09/manufacturing-offshore-is-bad-business
16. About outsourcing manufacturing to an offshore company, https://smallbusiness.chron.com/outsourcing-manufacturing-offshore-company-70663.html
17. Pros and cons: Foreign manufacturing or domestic? https://www.synthx.com/foreign-manufacturing-or-domestic/
18. S. Singh, N. Haldar, A. Bhattacharya, Offshore manufacturing contract design based on transfer price considering green tax: A bilevel programming approach. *Int J Prod Res* **56**(5), 1825–1849 (2018)
19. Why you should consider offshore manufacturing, August 2017, https://napsintl.com/mexico-manufacturing-news/consider-offshore-manufacturing/
20. 5 Cons of Offshoring, https://www.ruralsourcing.com/blogs/5-disadvantages-of-offshoring/?creative=469811773643&keyword=&matchtype=b&network=g&device=c&gclid=Cj0KCQjwxdSHBhCdARIsAG6zhlXBnPcK_KxgMgd-0YkWYfNFmlNLaf3vVhYvH1C1z6OEmb0maQyU7r0aAjNiEALw_wcB
21. J. M. Garcia, Offshoring in textile industry, January 2014, https://riunet.upv.es/bitstream/handle/10251/47923/Offshoring%20in%20textile%20industry%20project.pdf?sequence=1
22. Understanding offshore manufacturing. https://maverick-mcs.com/offshore-manufacturing-maquiladora/understanding-offshore-manufacturing/
23. Benefits of offshore manufacturing, https://www.sunfastusa.com/benefits-of-offshore-manufacturing
24. 5 Big problems for companies who offshore manufacturing, July 2017, https://www.shipstarter.com/single-post/2017/05/31/5-Big-Problems-for-Companies-Who-OffShore-Manufacturing
25. Outsourcing viewed as the top threat to U.S. jobs, https://www.statista.com/chart/6140/outsourcing-viewed-as-the-top-threat-to-us-jobs/

26. B. Bidanda, O. Arosoy, L.J. Shuman, Offshoring manufacturing: Implications for engineering jobs and education: A survey and case study. *Robot Comp Integra Manufact* **22**, 576–587 (2006)
27. E. Larki, The hidden costs of manufacturing offshore, *Machine Design,* February 7, 2008, p. 39.
28. S. Cherenfant (Compiler), *Domestic Vs. offshore manufacturing (current controversies) library binding.* Greenhaven Press, 2021.
29. A. Cokar, *Source my garment: The insider's guide to responsible offshore manufacturing* (Source My Garment Consulting Inc., 2019)
30. K. Thomsen, *Offshore wind: a comprehensive guide to successful offshore wind farm installation,* 2nd edn. (Academic Press, 2014)
31. E. Sakellos, *The use of offshore manufacturing and its impact on the U.S. industrial base* (New York Institute of Technology, 1991)
32. J.W. Schmits, *Offshore manufacturing: A study of u.s. manufacturers of consumer electrical products* (MIT Pres, 1980)
33. A. Asefeso, *Reshoring: manufacturing is coming home* (AA Global Sourcing Ltd., 2014)
34. C.D. Vance, *The effects of offshore manufacturing on the national* (University of Cincinnati, 2010)

Chapter 17
Cyber Manufacturing

Cybersecurity is a subject that requires logic, knowledge, thought, and commitment. It can be applied or research based. It is a true leveler for all to enter, be successful and lead the future of cybersecurity.

–Ian R. McAndrew

17.1 Introduction

Manufacturing embodies the process of creating a product out of raw materials. The manufacturing industry includes many sectors such as the food industry, chemical industry, automotive industry, aerospace industry, metal machinery industry, and electronics industry, to name a few. Manufacturing has been noted as one of the largest industries globally [1]. Manufacturing is a national priority in many nations, both in developed and developing nations. They are investing heavily in strategic, manufacturing-technology areas.

In the past, the manufacturing industry had been in a continuous struggle to produce high-quality goods at the reasonable price that consumers can afford. The industrial era has allowed machines to resolve the quantity issue of manufacturing. Many existing manufacturing systems were developed at a time when security was not an issue. Today's manufacturing operations integrate resources in a more complex manner, which requires extensive collaboration and security. Data (or information) is the biggest asset of a company, because it is how value is stored. This makes the security of the data to be a necessity.

Cybersecurity is one of the major hurdles in implementing a cyber manufacturing system (CMS), since CMS opens a door for cyber-physical attacks on manufacturing systems. Cyber manufacturing is a modern manufacturing approach that utilizes cyber-physical systems, which are smart systems that include co-engineered interacting networks of physical and computational components [2]. A cyber-physical system (CPS) is usually connected via the Internet. CPSs are essentially integrations of computation, networking, and physical processes. They are increasingly finding applications in manufacturing.

This chapter provides an introduction to cyber manufacturing. It begins by explaining what cyber manufacturing is all about. It describes enabling technologies

M. N. O. Sadiku et al., *Emerging Technologies in Manufacturing*, https://doi.org/10.1007/978-3-031-23156-8_17

needed in developing cyber manufacturing solutions. It presents common cyberattacks on manufacturing. It suggests some strategies for handling cybersecurity risks in manufacturing. It highlights some benefits and challenges of cyber manufacturing. It covers global cyber manufacturing. The last section concludes with comments.

17.2   What Is Cyber Manufacturing

The concept of "cyber manufacturing" is derived from cyber-physical systems. Cyber-physical systems (CPS) are engineered systems that are designed to interact seamlessly with networks of physical and computational components. They are integrations of computation, networking, and physical processes. They may be regarded as systems of collaborating computational elements that control physical entities, generally using feedback from sensors they monitor. They are complex systems with the integration of computation, communication, and control (3C) technology. They have made great strides in manufacturing systems. Common examples of CPS include automobiles, medical devices, and the smart grid (or power systems). Most CPS become safety-critical and security-critical.

Cyber manufacturing may be regarded as the convergence of cyber-physical systems (CPSs), systems engineering, and manufacturing innovation. CPSs are enabling technologies that bring the virtual and physical worlds together to create a truly networked world in which intelligent objects communicate and interact without human intervention. They are essentially the upgraded version of embedded systems. They are the heart of Industry 4.0. They are expected to bring about radical changes and revolution in the interaction between the physical and virtual worlds [3].

Cyber manufacturing is a new transformative concept that involves the translation of data from interconnected systems into predictive and prescriptive operations. This concept applies advancements in information technology to the manufacturing process. Expected functionalities of cyber manufacturing systems include machine connectivity and data acquisition and manufacturing reconfigurability.

Cyber manufacturing system (CMS) is a vision for future advanced manufacturing systems integrated with technologies. It offers a blueprint for future manufacturing systems in which physical components are fully integrated with computational processes in a connected environment. Figure 17.1 typically shows cyber manufacturing system [4].

As manufacturing becomes more digitized, it is becoming a popular target for cybercriminals, and manufacturing plant cybersecurity has become a hot issue. Manufacturers are often the target of cyberattacks because of their valuable data, positive value proposition, intellectual property (IP) assets, and trade secrets. They are increasingly vulnerable to attacks that can shut down production regardless of their size. Therefore, cybersecurity issues are gaining importance in the manufacturing industry. Cybersecurity is therefore concerned with a set of principles and means of ensuring the security of data. The stakes are high, and manufacturers should do everything possible in their power to alleviate the risks. Manufacturers must find ways to prevent attempts to corrupt their data, steal intellectual property, and sabotage their operation [5].

Fig. 17.1 Cyber manufacturing system [4]

17.3   Enabling Technologies

Several technologies are needed in developing cyber manufacturing solutions. The enabling technologies include cyber-physical systems (CPSs), Internet of Things (IoT), cloud computing, digital twin, sensors network, and machine learning.

- *Cyber-Physical System* (CPS) is the key foundation of cyber manufacturing.
- CPS usually refers to systems of collaborating computational elements that control physical entities. CPSs are based on connectivity but run complex analytics. They combine physical components and digital networks to change how manufacturing companies automate processes and information sharing.
- CPS stores and maintains data in the cloud using standard formats. In CPS, computation and networking technologies interact with physical systems. CPS technologies include wireless system integration, wireless controls, machine learning, and sensor-based manufacturing. CPSs are collaborating computational entities which are in intensive connection with the surrounding physical, providing and using data-accessing and data-processing services available on the Internet. They can be found in many applications including smart healthcare wearable devices, smart grid, smart water, smart manufacturing, smart factory, gas and oil pipelines monitoring and control, unmanned aerial and autonomous underwater vehicles, and hybrid electric vehicles. Fig. 17.2 illustrates the general representation of CPS [6].
- *Internet of Things* (IoT): Although there are similarities between CPS and IoT, they are not the same. The Internet of Things (IoT) is a network of physical

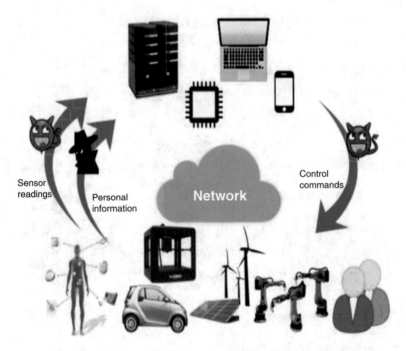

Fig. 17.2 General representation of CPS [6]

objects (sensors, machines, cars, buildings, etc.) that allows interaction and cooperation of these objects to reach a common objective. Wireless sensor networks (WSNs) are a key enabling technology for IoT in manufacturing. While the IoT paradigm does not include the idea of information analytics, CPS is based on connectivity and runs complex analytics. However, it is possible to view the IoT as the infrastructure that makes CPS possible. By increasing the amount of automation at multiple levels within a factory and across the enterprise, cyber-physical manufacturing systems enable higher productivity, higher quality, and lower costs. The real value of the IoT for manufacturers will be in the analytics arising from CPSs. Characteristically, IoT will make everything in our lives "smart." IoT has the potential to accurately track people, equipment, or even service animals and analyze the data captured [7].

- *Big Data Analytics*: The volume of data has exploded to unimaginable levels in the past decade. Big data comes from different sources such as sensors, devices, computer networks, machines, IoT, GPS, RFID, e-commerce transactions, web, weather data, medical data, insurance records, and social media. As we capture terabytes of data, our major challenge is making sense of this gigantic amount of data. This is where big data analytics fits into the picture. Big data analytics deals with examining massive amounts of data to uncover hidden patterns, market trends, customer preferences, and other useful information that can help in making informed decisions. The massive amount of raw data available from the manufacturing process creates opportunities to add intelligence to the process. Cyber manufacturing intertwines industrial big data and smart analytics to discover and comprehend invisible issues for decision-making. Big data analytics is a technology that aims to support manufacturers to extract knowledge from data, which can be used for future prediction, increasing profits, and enhancing customer service [8, 9].

17.4   Attack on Cyber Manufacturing

Cyberattacks in manufacturing industries are on the rise. Manufacturing remains one of the top sectors when it comes to targeted cyberattacks. The manufacturing industry is the fourth most vulnerable to cyberattacks, behind healthcare, government systems, and financial services. Automotive manufacturers are the top targets for criminals. Manufacturers are increasingly being targeted by traditional malicious actors such as hackers and cyber-criminals and by competing companies and nations engaged in corporate espionage. Motivations range from money and revenge to competitive advantage and strategic disruption. Manufacturing equipment have increasing vulnerabilities over the years.

It is important for businesses to be aware of some of the most common types of manufacturing cyberattacks so that you can take the necessary steps to protect your company. Types of cyberattacks in manufacturing vary from the illegal crime of an individual citizen (hacking) to the actions of groups (terrorists). The common attacks include the following [10–13].

- *Identity theft:* This is the most commonly reported form of cyberattacks. This kind of attack can either target the customer data or the data of employees. In manufacturing, there are countless incidents of malicious insiders stealing a company's intellectual property or other confidential information for personal profit or revenge. Intellectual property (IP) is the manufacturing industry's key asset and prized possession. The company's valuable IP can be stolen by a competing executive, causing significant reputational and financial damage to the company and loss in global operations. If a hacker can steal one-of-a-kind intellectual property, the hacker can sell the information for a premium.
- *Malware*: This is a malicious software or code that includes traditional computer viruses, computer worms, and Trojan horse programs. Malware can infiltrate your network through the Internet, downloads, attachments, email, social media, and other platforms. Spyware is a type of malware that collects information without the victim's knowledge. Advanced malware is another type of attack that is increasingly common in manufacturing. Internal threats can be just as damaging. To protect against malware, manufacturers can take the following steps:

  - File encryption.
  - Implement two-step verification.

- Install antivirus software on computers.
- Mandate encryption of data transmissions.
- Update operating systems, browsers, and plugins.
- Safeguard usernames and passwords.
- Secure Internet connections with firewalls.

- *Ransomware:* A ransomware attack locks users out of their computer systems and only allows access until an exorbitant amount of money (the ransom) is paid. This is now a well-known threat to manufacturing. Ransomware is a type of malware that encrypts files on a network and makes them unusable until the demands of hackers are met. These threat actors may threaten to sell or leak sensitive data if a ransom is not paid. Manufacturers have much to lose from ransomware because time is money for manufacturing companies.
- *Phishing*: This is a tactic where a criminal will seek out information and data on your business. Rather than stealing this, they will simply request the information. Phishing facilitates the process by tricking executives and their staff into revealing login credentials. A good example of a phishing scam is the "Nigerian Prince" who is ready to donate millions to you. All you need to do is provide your bank details and a copy of your passport. Criminals trick victims into handing over their personal information such as online passwords, social security number, and credit card numbers. Common warning signs of phishing include:

  - Attachments with dangerous file types
  - Immediate demands or requests
  - Lucrative offers with attention-grabbing statements
  - Misspelled hyperlinks of known websites
  - Unusual email sender(s)
  - Threats

- *Denial-of-service attacks*: These are designed to make a network resource unavailable to its intended users. These can prevent the user from accessing email, websites, online accounts, or other services. Manufacturing companies can fall victim to distributed denial of service (DDoS) style attacks. Recent research shows that manufacturing is the most targeted industry for DDoS.
- *Social Engineering Attacks*: A cybercriminal attempts to trick users to disclose sensitive information. A social engineer aims to convince a user through impersonation to disclose secrets such as passwords, card numbers, or social security number. A social engineer may send an e-mail saying, "Hey, this is vendor X, here's our new bank account information; please re-route all payments here."
- *Man-In-the-Middle Attack*: This is a cyberattack where a malicious attacker secretly inserts him/herself into a conversation between two parties who believe they are directly communicating with each other. A common example of man-in-the-middle attacks is eavesdropping. The goal of such an attack is to steal personal information.

Other cybersecurity threats that the manufacturing companies should be prepared for include IoT attacks, supply chain attack, internal breaches, equipment sabotage, and zero-day attacks. Cybersecurity involves reducing the risk of cyberattacks. It involves the collection of tools, policies, guidelines, risk management approaches, and best practices that can be used to protect the cyber environment and mitigate cyberattacks. Cybercrime prevention is a multifaceted issue. Cyber risks should be managed proactively by the management.

17.5 Handling Cybersecurity Risks

Every manufacturer is potentially at risk. Manufacturers are a major target for state-sponsored espionage as well as opportunistic hackers. There are strategic steps every manufacturer can take to reduce the security breach risks for their business and their customers. Any manufacturer of any size can handle cybersecurity risks by following these five steps [14].

1. *Identify:* Understand your resources and risks. Identify and control who has access to your business information. Manufacturers must "think outside of the box" to assess and identify cybersecurity threats. Effective cybersecurity is essentially knowing where the data lies.
2. *Protect:* If you experience a cyberattack, you need to be prepared to resist. Make sure that the right security systems are in place. Guard against both targeted and untargeted attacks. Limit employee access to data and information. Enforce password rules requiring strong passwords and two-factor authentication when available. Train your employees so they can recognize a potential breach or attack. For some manufacturers, the biggest cyber threat is a lack of knowledge and training of the employees.
3. *Detect:* If an attack occurs, you need mechanisms in place that will alert you as quickly as possible. Install and update your systems with the latest antivirus software. Also, update your cybersecurity plan from time to time.
4. *Respond:* If a cybersecurity breach happens, you need to contain and reduce any damage. Stay updated on potential cyber threats.
5. *Recover:* After a cybersecurity breach, you need mechanisms in place to help resume normal operations. Make full backups of important business data and information.

The five steps are illustrated in Fig. 17.3 [14]. These strategies should be applied to mitigate risk and secure key assets. Unfortunately, there is no silver bullet to ensure complete security for manufacturers. To secure manufacturing systems, it is important to invest in both preventative measures and active defense including: cryptographic countermeasures, systems of intrusion detection, and proactive staff training.

Fig. 17.3 Five steps a
manufacturer can take to
handle cybersecurity
risks [14]

17.6 Benefits

Most manufacturing companies are behind in cybersecurity systems implementation. Some cyberattacks on manufacturing impact operations so severely that they
can pause operations, close plants, precipitate companies reneging on customer
orders, and cause companies to spend substantial sums to recover from the event.
Strong cybersecurity is the foundation for a resilient manufacturing company.
Unfortunately, there is no silver bullet to ensure complete security for manufacturing companies. Food manufacturers always look for ways to prevent cyberattacks
from keeping their businesses and the food products they produce safe for consumers.

Trends, such as the industrial Internet of Things and Industry 4.0, are driving
organizations to facilitate more connections between the physical process world and
the Internet. The risk to manufacturing organizations from cyberattacks has
increased dramatically. However, there are strategies to mitigate risk and secure key
assets. To be ready for the future, manufacturing companies must be prepared to act
now. It is always expedient to follow a good old principle: hope for the best, plan for
the worst. As IoT and CPS technologies advance, cyber manufacturing realization
will be facilitated through standardization and more reliable cybersecurity.

17.7 Challenges

The applications of CPS in many domains such as manufacturing will have disruptive effects on technology, business, law, and ethics. The increased use of CPS
brings some threats that could have significant consequences for users. CPS are

prone to information leakage from the physical domain. They are prone to a wider range of attacks and design flaws [15].

Cyber-physical attacks are becoming common across a wide range of industries including manufacturing. Manufacturing is a significant target for cybercriminals, and manufacturers must take appropriate actions to protect themselves in the face of cyber-related risks. The attacks may cause physical damages to physical components (such as machines, equipment, parts, products, etc.) through unintended over-wearing, breakage, and scrap parts. Cyber-physical attacks are new and unique risks to CMSs. The four most dangerous cyber risks are identity theft, compromised websites, spam, and employees [16].

Although there is no mandated regulatory standards for governing cybersecurity in manufacturing, the cybersecurity communities have developed a variety of standards, which address weaknesses and vulnerabilities. Different standards contribute in different ways to realize cyber manufacturing. Developing standards tends to motivate the adoption of new technologies and accelerates the adoption of new manufactured products and manufacturing methods. Standards allow a more dynamic and competitive marketplace. They are a critical tool for leveling the playing field for small businesses by reducing the cost barriers. They enable low-cost applications which are suitable for small manufacturers [17].

These challenges are critical to cyber manufacturers' abilities to capture the value associated with this new frontier of technology while appropriately addressing the dynamic cyber risks. Many technologies, such as big data analytics, communication protocols, and cybersecurity, still need a lot of research and development before cyber manufacturing can reach their full potential [18].

17.8 Global Cyber Manufacturing

In today's business environment of increased automation, ubiquitous connectivity, and globalization, even the most powerful organizations in the world are vulnerable to debilitating cyber threats. As manufacturing around the world are implementing digital transformation to survive, grow, and drive smart factories, cybersecurity threats have become a major concern. The manufacturing industry is one of the most targeted sectors globally due to its vulnerability. US and European countries still can lead international manufacturing by exploiting cyber physical system (CPS) technologies such as wireless system integration, wireless controls, machine learning, and sensor-based manufacturing. We now consider how cyber manufacturing is being applied in some nations [19, 20].

- *United States:* The US House of Representatives approved three (3) bills that aim at strengthening cybersecurity measures in critical energy infrastructure. It also took concrete measures to bolster industrial control cybersecurity, strengthen US critical supply chains, and improve long-term economic security. The US Department of Energy (DOE) released the Cybersecurity Capability Maturity

Model to meet cyber threat challenges. The US Department of Justice has charged five (5) Chinese military officers with stealing trade secrets from manufacturers such as Westinghouse Electric, US Steel, Alcoa, and Allegheny Technologies. These moves demonstrate that US government agencies are taking the threat against the security of the cyber-physical systems very seriously.

- *United Kingdom:* Cyber threat has been the biggest risk for manufacturing in the United Kingdom. Most manufacturers have not received training to deal with cyberattacks and some are largely unprepared for the growing possibility of a cyberattack. Manufacturers, particularly hi-tech manufacturers, such as aerospace, electronics, and pharmaceuticals, were a high-risk target. As the UK's voice for manufacturing and engineering, European Energy Forum (EEF) has the potential to play a significant role in supporting manufacturers in the face of some challenges. Effective cybersecurity at both the national level, and for businesses, requires cooperation between the public and private sectors. The individual forces are supported by the Regional Organized Crime Units (ROCUs).
- *Germany:* In 2014, a blast furnace at a German steel mill reportedly suffered "massive damage" after hackers used malware-laden emails to gain access to the unnamed steel mill's automated control systems. While the exact details of the company involved are still unknown, the attacker used sophisticated social engineering and phishing tactics to hack into the German steel mill's computer network. The attacker, likely an industry insider or someone working with an insider, had specific knowledge of the production processes involved so that maximum damage could be done.

17.9 Conclusion

In an era of ubiquitous connectivity when more and more industrial systems are connected to the Internet, manufacturing is a prime target for cyber criminals. As manufacturing increasingly becomes more digitized, cybersecurity must become a critical standard component of effective management in manufacturing. Cybersecurity is now becoming a dominant concern for manufacturers and consumers. By increasing the level of automation within a factory, cyber-physical manufacturing systems enable higher productivity and higher quality at lower costs.

The most important part of any manufacturing operation is its people. To help in the attraction of cyber talent, manufacturers should work to become allies with higher educational institutions and build the talent pool they need. While attracting and retaining talented cyber professionals, manufacturers should provide training for the rest of the company's employees. More information about cyber manufacturing can be found in the books in [21, 22] and related journals: *Manufacturing Letters* and *Journal of Manufacturing Systems*

References

1. M.N.O. Sadiku, T.J. Ashaolu, A. Ajayi-Majebi, S.M. Musa, Emerging technologies in manu-facturing. Int J Scientif Adv **1**(2) (2020)
2. M.N.O. Sadiku, P.O. Adebo, A. Ajayi-Majebi, S.M. Musa, Cyber manufacturing. Int J Eng Res Technol **10**(4), 168–171 (2021)
3. A.R. Al-Ali, R. Gupta, A.A. Nabulsi, *Cyber physical systems role in manufacturing technologies* (AIP Conf Proc, 2018). https://doi.org/10.1063/1.5034337
4. J. Lee, B. Bagheri, C. Jin, Introduction to cyber manufacturing. Manufact Lett **8**, 11–15 (2016)
5. How to strengthen cyber security in smart manufacturing, September 2020, https://www.aeo-logic.com/blog/how-to-strengthen-cyber-security-in-smart-manufacturing/
6. M.N.O. Sadiku, K.G. Eze, S.M. Musa, Cyber-physical systems: A brief survey. *Int J Trend Res Develop* **7**(3) (2020)
7. M.N.O. Sadiku, S.M. Musa, S.R. Nelatury, Internet of things: An introduction. Int J Eng Res Adv Technol **2**(3), 39–43 (2016)
8. M.N.O. Sadiku, J. Foreman, S.M. Musa, Big data analytics: A primer. Int J Technol Manag Res **5**(9), 44–49 (2018)
9. C.M.M. Kotteti, M.N.O. Sadiku, S.M. Musa, Big data analytics. Invent J Res Technol Eng Manag **2**(10), 2455–3689 (Oct. 2018)
10. Global cyber executive briefing, https://www2.deloitte.com/global/en/pages/risk/cyber-strategic-risk/articles/Manufacturing.html
11. J. Miller, Popular types of cyber attacks in manufacturing, August 2019, https://www.bitlyft.com/resources/popular-types-of-cyber-attacks-in-manufacturing
12. M.N.O. Sadiku, S. Alam, S.M. Musa, A primer on cybersecurity. *Int J Adv Scientif Res Eng* **3**(8), 71–74 (2017)
13. B. Schmidt, 3 types of cyber attacks your manufacturing business needs to anticipate, April 2019, https://www.manufacturingsuccess.org/blog/3-types-of-cyber-attacks-your-manufacturing-business-needs-to-anticipate
14. E. Forsyth, MAMTC newsletter – 9/18. Cybersecurity. Steps manufacturers need to know, https://www.wearekms.com/mamtc-newsletter-918-cybersecurity-steps-manufacturers-need-to-know
15. M.N.O. Sadiku, Y. Wang, S. Cui, S.M. Musa, Cyber-physical systems: a primer. Eur Sci J **13**(36), 52–58 (2017)
16. G. Iltis, The 4 most dangerous manufacturing cybersecurity threats, January 2020, https://www.tech-pointe.com/blog/manufacturing-cybersecurity-threats
17. A. B. Feeney, S. Frechette, and V. Srinivasan, Cyber-physical systems engineering for manufacturing, in *Springer Series in Wireless Technology, Industrial Internet of Things: Cybermanufacturing Systems,* chapter 4, 2017
18. Deloitte, Cyber risk in advanced manufacturing, https://www2.deloitte.com/content/dam/Deloitte/us/Documents/manufacturing/us-manu-cyber-risk-in-advanced-manufacturing.pdf
19. Greater cyber resilience necessary to cope with security risks in manufacturing and smart factories, March 2021., https://industrialcyber.co/article/greater-cyber-resilience-necessary-to-cope-with-security-risks-in-manufacturing-and-smart-factories/
20. Cyber security for manufacturing - make uk, https://www.makeuk.org
21. L. Wang, X.V. Wang, *Cloud-based cyber-physical systems in manufacturing* (Springer, 2018)
22. S. Jeschke et al., *Industrial Internet of Things: Cybermanufacturing Systems* (Springer, 2017)

Chapter 18
Biomanufacturing

I think the biggest innovations of the twenty-first century will be at the intersection of biology and technology. A new era is beginning.

–Steve Jobs

18.1   Introduction

Manufacturing constitutes a significant portion of the economy of any nation and has a rich history. We are currently experiencing a major change in manufacturing, the means of production. The biomanufacturing revolution is changing the way we think of centralized systems, social equity, and our relationship with nature. It is becoming a change agent similar in scope to the Industrial Revolution [1].

Biomanufacturing (or biologics manufacturing) is the production of biological products from living cells. It may be regarded as the manufacturing component of the biotechnology industry. It is an emerging manufacturing process that utilizes biological systems to produce commercially important biomaterials and biomolecules. It is a highly complex process that requires much more time and expense than needed for small molecules. It is used in medicine, food industry, and industrial applications. Biomanufactured products range from biopharmaceuticals to industrial enzymes, human tissues, and replacement organs. They are used across several industries such as food and beverage processing and for industrial applications.

These products are designed to improve lives, fight disease, save lives, boost the economy, increase food production, and make it possible to feed the entire world. Fig. 18.1 illustrates a typical biomanufacturing system [2].

This chapter introduces the reader to biomanufacturing. It begins by describing what biomanufacturing is all about. It provides some applications of biomanufacturing. It highlights the benefits and challenges of biomanufacturing. It covers global biomanufacturing and the future of biomanufacturing. The last section concludes with comments.

© The Author(s), under exclusive license to Springer Nature Switzerland AG 2023
M. N. O. Sadiku et al., *Emerging Technologies in Manufacturing*,
https://doi.org/10.1007/978-3-031-23156-8_18

Fig. 18.1 A typical biomanufacturing center [2]

18.2   What Is Biomanufacturing?

Biotechnology is essentially technology based on biology. It is the use of advanced manufacturing approaches to produce the next generation of healthcare products. It is the use of living organisms and biological processes to solve problems or make useful products. Biomanufacturing, a branch of biotechnology, is an advanced-technology manufacturing that is responsible for making biopharmaceuticals. It can be traced back to the 1860s, when fermentation products appeared on the market. Biomanufacturing is multidisciplinary covering a wide range of fields including biochemistry, computation, process engineering, microbiology, synthetic biology, regenerative medicine, tissue engineering, mechanical engineering, robotics, software, and bioprinting [3]. Figure 18.2 shows a typical world-class biomanufacturing facility in California [4]. (California is a manufacturing giant, containing 13% of the total manufacturing companies in the United States.)

Biomanufacturing (or biotech manufacturing) involves engineering a cell to produce a specific protein. It is being fueled by nature-based tools such as fermentation, enzymes, and microorganisms. Once a manufacturer successfully manipulates a cell to produce said protein, the cells multiply. The two popular production cells are *E. coli* cells and GFP cells, shown Fig. 18.3 [5]. For example, the first biologic drug, insulin, was produced using *E. coli* cells. The manufacturer establishes a master cell bank for future products. The cells are left to multiply for a few generations, creating hundreds of millions of identical copies. To avoid losing the cells in case of disaster, the manufacturer divides the master cell bank for storage into three separate locations. A cell bank production is shown in Fig. 18.4 [6].

Fig. 18.2 A typical world class biomanufacturing facility in California [4]

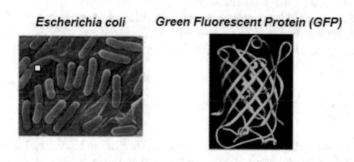

Fig. 18.3 *E. coli* and *E.coli* GFP [5]

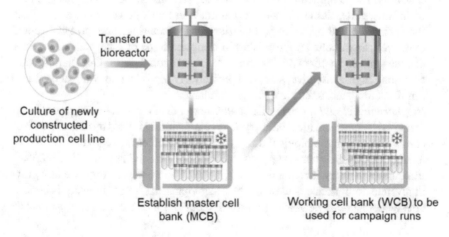

Fig. 18.4 A cell bank production [6]

The most common methods of biomanufacturing useful cells include the following [7]:

- *Blood Plasma Fractionation:* This is the process of separating components of blood plasma, which are most commonly used in medicines to improve the immune system.
- *Cell Culture: This* is the process by which cells are grown in an artificial environment under controlled conditions. For example, animal and plant cells are replicated and grown in a culture dish. This is beneficial for medicines, supplements, and enzymes.
- *Column Chromatography: This* method is used to purify chemical compounds from mixtures. For example, the method is used for pharmaceuticals and medicines.
- *Fermentation:* This is a metabolic process converting sugar into gases, acids, and alcohol. It is most important for the food and beverage industry as well as for industrial applications. Fermentation makes use of microorganisms and enzymes to carry out reactions. It is the source of some common products such as bread, beer, soaps, detergents, alcohol, and more.
- *Homogenization: This is* the process by which all fractions in a biological sample are equally made. This is useful in medicine and industrial applications to create a balance among cell constructions.
- *Ultrafiltration: This* process involves the separation of solids and solutes. Low molecular weight components pass through the semipermeable membrane. The process is used in drinking water, wastewater treatment, etc.

18.3 Applications

Biomanufacturing is used to produce commercially relevant biomaterials to be used in medicine, industrial applications, and the food and beverage industry.

- *Medicine:* Biomanufacturing uses biological systems including enzymes, microorganisms, cells, tissues, plants, and animals to fabricate commercial products for agricultural, food, energy, materials, manufacturing, and pharmaceutical industries. Stem cells are promising to bring the hope of a permanent cure for diseases that currently cannot be cured by conventional drugs. Cell therapies are becoming highly effective, safe, and predictively reproducible while at the same time becoming affordable and widely available.
- *Biopharmaceutical Industry:* The biopharmaceutical market is an important segment of the pharmaceutical industry. Biomanufacturing is used by pharmaceutical companies to make biological drugs such as antibodies and enzymes for bioremediation. (As their name suggests, biologics require living cells.) The biopharmaceutical industry is dynamic and complex. It continues to improve and grow in revenue, importance, and diversity. The biopharmaceutical industry is continuously demanding new and improved bioprocessing technologies to reduce

Fig. 18.5 Biomanufacturing technician [8]

costs, increase efficiencies, and improve weak development pipelines, especially in developing economies. Figure 18.5 shows a biomanufacturing technician [8]. Demand forecasting is an important factor that influences pharmaceutical and biopharmaceutical manufacturing decisions. Accurate forecasting involves cost-saving decisions on taxes, intellectual property strategy, and insurance of capacity. It influences decisions regarding capital cost, outsourcing for product commercialization, and other aspects of production.

- *Stem Cell Biomanufacturing:* Most stem cell-associated biomanufacturing is limited to the fabrication of tissue engineering devices. Stem cells outstand in future therapeutic applications. Stem cells are gaining attention due to their self-renewal ability and capacity of differentiating into any specific cell type [3]. Stem cells are providing hope for a permanent cure for diseases and disorders that currently cannot be cured by conventional drugs. This paradigm shift in modern medicine of using cells as novel therapeutics can be realized only if suitable manufacturing technologies for producing high-quality cells are available [9]. Fig. 18.6 depicts some emerging technologies for cell and tissue biomanufacturing [10].

- *Tissue Engineering:* This is often referred to as regenerative medicine and reparative medicine. It has emerged as an interdisciplinary field that combines the efforts of cell biologists, engineers, material scientists, mathematicians, geneticists, and clinicians toward the development of biological substitutes that improve tissue function. Cells placed within constructs are the most common strategy in tissue engineering. In tissue engineering, additive fabrication processes have been used to produce scaffolds with customized external shape [11].

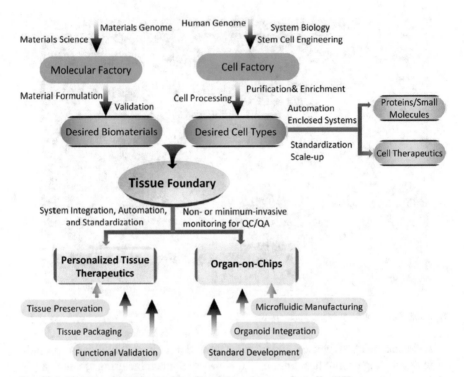

Fig. 18.6 Emerging technologies for cell and tissue biomanufacturing [10]

Tissue engineering is a reliable option to cure cardiovascular disease, which is a leading cause of death in the world. Tissue engineering has also been employed to promote bone and cartilage regeneration. Biomaterials play an important role in tissue engineering and regenerative medicine. With the increasing demand for tissue repair, a plethora of biomaterials are modified and exploited in various applications.

- *Food Manufacturing:* The food processing industry uses biomanufactured products. The fermentation process (enzyme and microorganism reaction) has found an ever-increasing suite of uses throughout food manufacturing. Products manufactured for food and beverage must be produced in a facility that has been designed within the confines of good manufacturing practices (GMP) regulations.

- *Additive Biomanufacturing*: This is perhaps the most exciting potential application of biomanufacturing. It produces on-demand tissues and even organs through a "3D printing." platform. This is the use of 3D printing. Additive manufacturing has been used to produce implants specially designed for a particular patient, with sizes, shapes, and optimized mechanical properties in many areas of medicine. 3D printing technologies are potential techniques for the development of porous mechanically tuned metallic and ceramic-based implants for medical applications. They allow the design and fabrication of a range of products from preoperative models, cutting guides, and medical devices to scaffolds. Additive manufacturing has been used to produce implants specially designed for a particular patient, with sizes, shapes, and mechanical properties optimized [12].

- *Smart Biomanufacturing:* To improve yield, reduce time to market, and facilitate process validation, smart biomanufacturing processes must deliver insights into biological behavior. The race for smart biopharma processes is driven by three trends: personalized drugs, flexibility as a means to make plants smarter, and using digitalization to acquire a deep process understanding for continuous, real-time quality control [13].
- *Continuous Biomanufacturing:* This refers to processes that occur on an ongoing basis after an initial catalyzation, with a minimum of monitoring. It requires continuous, accurate process monitoring. The trend toward improving process efficiency and control has advanced continuous manufacturing processes. The benefits of continuous manufacturing include significant economic advantages, reduced equipment size, enhanced cost efficiency, and high volumetric productivity. As a result, there is a lot of interest in small-scale continuous manufacturing technology although this is challenging. Challenges such as population heterogeneity and loss of the catalyst need to be overcome. Continuous bioprocessing (both in upstream and downstream operations) has been proposed as one future biomanufacturing state because it operates at similar scales for both clinical and commercial production. It requires working with multiple smaller columns. Through continuous bioprocessing, large amounts of the product may be manufactured with smaller equipment, decreasing the overall capital investment. Continuous bioprocessing can result in better-quality products due to higher purities and lesser impurities. Continuous biomanufacturing potentially provides significant improvements in the development and production of biopharmaceuticals. The drug industry has been slow to adopt continuous biomanufacturing largely because of costs and regulatory concerns [14].
- *Electronics:* Biomanufacturing presents exciting possibilities in electronic component manufacturing such as printed circuits and touch sensors at a fraction of the energy cost of traditional materials.
- *Consumer Products*: Biomanufacturing produces all manner of consumer goods including beauty supplies, plastic products and components, nylon, textiles, paper, and more.
- *Single-Use Systems*: Since their initial development, single-use systems are being increasingly used in almost all stages of biomanufacturing processes. They are new technologies that dramatically lower building of biomanufacturing plants. Their intention is to create factories that can manufacture multiple products at a single site in a modular manner.

Other areas of applications of biomanufacturing include perfusion biomanufacturing, cell-free biomanufacturing, plant-based protein biomanufacturing, commercial biomanufacturing, cell line development, and the military.

18.4   Benefits

Biomanufacturing is an industrial production technology that uses living cells as miniature factories to produce products designed to detect, modify, maintain, and study biological organisms.

Biomanufacturing also provides opportunities for multidisciplinary collaborations, both in academia and industry. Challenges facing humankind, such as food security, renewable energy, urbanization, and climate change motivate rapid developments in biomanufacturing platforms. Much interest in biomanufacturing relates to medical products, but agriculture and livestock farming are also biomanufacturing at their most fundamental. Other benefits of biomanufacturing include [7, 15].

- *Biomanufactured Products:* Thousands of biomanufactured products are available on the market today. They are found in natural sources like cultures of microbes, blood, or plant and animal cells. They are classified by medicine, food and beverage, and industrial applications. In the medical field, amino acids, vaccines, cytokines, fusion proteins, growth factors, biopharmaceuticals, and monoclonal antibodies all utilize biomanufactured products. In the food and beverage industry, biomanufactured products include amino acids, enzymes, and protein supplements. Industrial applications that require biomanufactured products include biocementation, detergents, plastics, and bioremediation.
- *Energy Use Reduction:* Biomanufacturing vastly reduces this energy requirement in favor of production through natural reactions. It replaces energy-intensive processes with naturally occurring processes that often require only sunlight and a controlled environment to be initiated. Lower energy usage and costs benefit the environment, as well as manufacturers and consumers.
- *Increased Innovation:* It is already impacting a wide range of industries, and it is driving incredible, unprecedented technological advancements. The possibilities of biomanufacturing are providing our best minds with the ideal jumping-off point to innovate across numerous industries.
- *Sustainability:* Materials and products created through biotechnology are easier and safer to recycle and dispose. The long-term sustainability of biomanufacturing can solve the problems of waste disposal and toxic material. Growing biomanufacturing will help enhance a sustainable economy and green biomanufacturing.

18.5   Challenges

In manufacturing, interactions between humans and machines are changing, and machines are acquiring decision-making capabilities. The biomanufacturing industry is no exception. It is important to understand this and balance capacity versus demand and staffing level. Manufacturing medicine using biology poses different challenges from the traditional chemical manufacturing processes. Significant improvements in biomanufacturing are required to enhance safety, quality, and consistency of biopharmaceutical products while reducing the manufacturing cost. As with any emerging technology, the biggest challenge is changing the current mindset. New entrants to emerging markets will face many challenges and opportunities, depending on the particulars of the location they target.

- *Complexity:* Biomanufacturing is a complex, labor-intensive industry requiring a combination of skills from several disciplines. The inherent complexity increases the drag on the process of innovation. Biomanufacturing has been shaped by large capital requirements, high operating costs, and complex processes. This has made it an expensive business to operate.
- *Affordability:* Commercial biomanufacturing faces a major challenge as cost elements continue to be a huge problem. Establishing a new commercial biomanufacturing facility may cost one billion dollar. The affordability and cost-effectiveness of biologic therapy are increasingly important. The cost of goods of biologics is still relatively high. We need to find innovative ways of reducing this to ensure these types of biological therapies and vaccines are more affordable. The pharmaceutical industry is at an important crossroad. It must cut costs and develop safe and cost-efficient processes for new biological products.
- *Hiring Problems*: As the bioprocess industry expands, some facilities will face major hiring problems. The apparent decrease in hiring problems may be due to the fact that some companies have been successful in streamlining their processes, making them more efficient, more automated, faster, and less expensive. That can reduce pressure on workers [16].
- *No Regulations*: Regulation is a friend, not a foe of innovation, in bioproduction processes. The industry has no regulatory requirements for single-use technologies (SUTs). Some specific single-use parts are listed in the requirements, but such requirements are not defined for all production steps [17]. This lack of regulation is only at the local level. However, biomanufacturing processes is heavily regulated globally, due to its use of genetically modified organisms.
- *Lack of Infrastructure*: The lack of appropriate biomanufacturing and supply-chain infrastructure is a key barrier for widespread and equitable distribution of vaccines for COVID-19 across the world. This requires a fundamental change in national and global science policy.
- *Workforce Challenges*: These including balancing collaboration with competition for talent, expanding the talent pipeline, and securing the funding required to support essential workforce development. Competition among companies vying to hire the best talent creates an inherent obstacle to executing a collaborative regional workforce strategy. What is needed in manufacturing is building both executive-level talent and the technical biomanufacturing workforce. To achieve this, the region must strike a balance between recruiting competition and executing a cohesive, collaborative regional talent strategy [18].

Figure 18.7 illustrates these challenges.

Fig. 18.7 Some challenges of biomanufacturing

18.6   Global Biomanufacturing

Biomanufacturing is beginning a global transformation. It produces a wide range of products for the global bioeconomy. The changes taking shape in the global biomanufacturing industry will fundamentally alter the global playing field. Success in the globally distributed biomanufacturing industry will be a function of how well companies can leverage new technologies and manage the challenges of biomanufacturing. A lion's share of global biomanufacturers are located in North America, Western Europe, Ireland, and the United Kingdom. We now consider how biomanufacturing is being implemented in some nations.

- *United States:* The United States is the world's leading producer of manufactured goods. It is also a global leader in biomanufacturing, and the industry continues to grow in US biomanufacturing as an industry is experiencing explosive growth within the United States. It is a manufacturing revolution that can make the United States self-sufficient. American companies are working to build a sustainable, domestic manufacturing ecosystem that will enable domestic bio-industrial manufacturing, develop technologies to enhance US bio-industrial competitiveness, de-risk investment in relevant infrastructure, and expand the biomanufacturing workforce to realize the economic promise of industrial bio-technology. Although the United States leads the world in innovations and inventions, the manufacturing capabilities and new products are often "stranded in the lab," or get outsourced to other nations. The competitiveness of US manufacturing has been challenged by other developed nations such as Japan, Germany, and Korea, as well as developing nations such as China [19].

- *United Kingdom:* The chemical industry is vitally important to the UK economy. The United Kingdom has a solid research base from which to grow a vibrant biomanufacturing economy. The United Kingdom must continue to build on excellence if it is to remain a prime bioprocessing location. What the United Kingdom is doing right has been assessed and recommendations have been made regarding, and the direction it needs to head in the coming decade if it is to stay a leading center for biomanufacturing. The UK has become a biomanufacturing cluster similar to Research Triangle Park or Boston in the United States with 260 companies, and around 12 thousand people actively focused on developing or manufacturing biologicals. There are rumors that the United Kingdom intends to remain a top location for the production of biologicals and is going to increase its biomanufacturing capacity [20]. The United Kingdom cannot do everything and needs to be strategic. The nation could focus on manufacturing value-added products and develop the innovation expertise to export processes and technologies to manufacturing organizations overseas [21].

- *China:* Biomanufacturing is relatively new in China; the country is just getting started. An emerging trend in the contract manufacturing organization (CMO) market is the rise of China as both a customer and potential supplier of biologics. As biopharmaceutical growth is also increasing in China. China, as a high-growth region, is moving at unprecedented speeds and is thirsty to train its workforce in biologics GMP best practices to meet the demand for skilled labor. China is likely to be America's chief competitor in what may prove to be a super-power manufacturing technology race in which staying ahead will be a national security imperative.

- *Australia:* An Australia-based contract manufacturing organization, Luina Bio, offers microbial fermentation services for drug development and manufacturing. The company has more than 20 years of experience and produce commercial veterinary products for clients within Australia and internationally. Luina Bio's manufacturing facility houses fermentation vessels and offers a full suite of manufacturing for both whole-cell preparations. The company offers world-leading fermentation and related services in fermentation and subsequent downstream purification [22].

- *Canada:* The government of Canada ensures access to critical vaccines, therapeutics, and other health products. At the outset of the COVID-19 pandemic, the Government of Canada took early and decisive action to strengthen Canada's biomanufacturing capacity to protect Canadians. The Government of Canada has also committed a substantial amount of money to support research and development, clinical trials, and future pandemic preparedness [23].

- *Russia:* The Russian company R-Pharm is a good example of the evolving bio-manufacturing industry in research and development. The success of its distribution business has enabled the company's strategic plan to build a fully integrated drug discovery, development, manufacturing, and distribution across Russia. R-Pharm is well an established, successful business employing top talent. It has access to excellent science, both in-house and through partnerships with leading universities. R-Pharm works with domestic and international partners like the United States and China to simultaneously advance the core elements of its biologics program. This strategy is designed to significantly reduce time to clinic and market [24].
- *South Korea:* The South Korean biopharmaceutical company Samsung Biologics is set to break ground as one of the world's largest manufacturing facilities within a single location. It also helps streamline production. The super biomanufacturing facility will be both larger and more expensive than all three of Samsung's existing plants combined. The global scramble for biomanufacturing, drug development resources, and COVID-19 treatments prompted the extreme increase in size. The ambitious facility will begin taking on outsourced manufacturing demand due to COVID-19 [25].
- *Singapore*: Biomanufacturing constitutes a critical pillar in Singapore's economy and has much potential to be further strengthened through innovation. With strong expertise in digitalization and manufacturing, Singapore can strategically leverage its capabilities to drive the local industry for the future by addressing resource sustainability, diversification of supply and logistic chains, and affordable drugs. With the guiding objective of promoting research and development, feasible breakthroughs in biologics manufacturing technologies can increase operational efficiencies. Talent development through training remains the main thrust, as competent and skilled workforce is needed [26].

18.7   Future of Biomanufacturing

The dynamics of biopharmaceutical manufacturing are changing. Biopharmaceutical companies are booming and growing at an exponential rate due to the robust adoption of advanced technologies in the biomanufacturing industry such as artificial intelligence, augmented reality, and digital twins. The companies continue to expand in revenue, breadth, importance, and diversity. Globalization is increasing the industry's geographic scale, and clearly the pharmaceutical market is continuing to expand its reach globally. By exploring the history of the biomanufacturing market and by analyzing both past and ongoing key trends, one can predict what the future of the industry is likely to hold. We identify and outline key production trends below and build toward a detailed vision for the future of biomanufacturing. By harnessing the best aspects of biomanufacturing, we can create a future where a new economy supports local communities, empowers new innovation, and spreads economic power to segments of society left behind.

The key technology trends that are influencing the biomanufacturing market include [27]:

- *Cost reduction:* Due to the high cost associated with the biomanufacturing industry, the requirement to reduce cost is imperative.
- *Collaboration:* Biomanufacturing provides great opportunities for multidisciplinary collaborations. Different manufacturing sectors will need to share ideas and technology. Collaboration between research and development teams and manufacturing partners is an important factor in future biomanufacturing. Their effort must focus on the next-generation biomanufacturing.
- *Advanced Technology*: Advanced manufacturing technologies can propel innovations. Adoption of advanced technologies as well as new innovation in the form of design, efficiency, performance, and others are some of the factors that will likely enhance the growth of the next-generation biomanufacturing market, which will boost various lucrative opportunities. The next-generation biomanufacturing market is expected to gain market growth.
- *Increased Funding:* Biomanufacturing growth will be dictated by price controls imposed directly by governments and the prevalence of funds from government and from private organizations for the development of innovative technology which will further boost various opportunities.
- *Automation:* Automated systems are gaining a foothold in the market as they can reduce costs related to the production process and can also enhance productivity. Automation also reduces manual, human errors.
- *Single-Use Technologies:* These are gaining widespread acceptance with more preference for systems that are connected in a sequence and are being used for large-scale manufacturing.

These biomanufacturing trends are appropriate for regional communities and promise to leave our environment and society more adaptable and resilient.

18.8   Conclusion

In essence, biomanufacturing applies the principles of engineering and chemical design to biological systems. Biomanufacturing is the use of advanced manufacturing approaches to produce the next generation of healthcare products. The biomanufacturing revolution is changing the way we think of centralized systems and our relationship to nature.

Although biomanufacturing has played an important role in the past three industrial revolutions, biomanufacturing 4.0 (or advanced biomanufacturing) will be one of the most important cornerstones of the sustainability revolution of the twenty-first century [28].

Biomanufacturing academic programs are designed to prepare students for positions in pharmaceutical manufacturing, bioprocessing, media preparation, bioenergy production, and the food processing industry. Well-informed and prepared graduates can better manage their own career, understand job requirements, and get their "foot in the door." Biomanufacturing jobs run the gamut from bachelor's degree level positions to PhD creative roles that perform development and research. The rapid growth of manufacturing operations has created jobs in different areas such as science (e.g., biochemistry, chemistry, or microbiology) and engineering (chemical engineering, mechanical engineering, biochemical engineering) [29]. For more information about biomanufacturing, the reader should consult the books in [14, 30–38] and related journals:

- *Manufacturing Letters*
- *Biotechnology Annual Review*
- *Biotechnology Letters*
- *Biotechnology Entrepreneurship*
- *Journal of Industrial Microbiology and Biotechnology*
- *Frontiers in Bioengineering and Biotechnology*

References

1. R. Lipski, The biomanufacturing revolution is here and it will change the world, May 2020, https://medium.com/predict/the-biomanufacturing-revolution-is-here-and-it-will-change-the-world-7f431814fcb3#:~:text=The%20Biomanufacturing%20Revolution%20is%20Here%20and%20it%20Will%20Change%20the%20World&text=means%20of%20production.-,The%20biomanufacturing%20revolution%20is%20changing%20the%20way%20we%20think%20of,use%20in%20our%20daily%20lives
2. COVID-19 therapeutics to rely on large-scale biomanufacturing reactors, report says, https://www.europeanpharmaceuticalreview.com/news/133022/covid-19-therapeutics-to-rely-on-large-scale-biomanufacturing-reactors-report-says/
3. Y. Xu et al., Biomaterials for stem cell engineering and biomanufacturing. Bioactive Mater **4**, 366–379 (2019)
4. Vacaville unveils 'California Biomanufacturing Center' plan to spur $2B in building, 10,000 jobs, https://www.northbaybusinessjournal.com/article/article/vacaville-unveils-california-biomanufacturing-center-plan-to-spur-2b-in/
5. Biopharmaceutical manufacturing (biomanufacturing), Unknown Source
6. E. Burke, Biomanufacturing: how biologics are made, June 2020, https://weekly.biotechprimer.com/biomanufacturing-how-biologics-are-made/
7. G. Parker, What is biomanufacturing and how will it change the world? https://moneyinc.com/biomanufacturing/
8. Improving biomanufacturing with automation, April 2019, https://industryeurope.com/improving-biomanufacturing-with-automation/
9. K. Roh, R.M. Nerem, K. Roy, Biomanufacturing of therapeutic cells: state of the art, current challenges, and future perspectives. Ann Rev Chem Biomol Eng **7**, 455–478 (2016)
10. K. Ye et al., Advanced cell and tissue biomanufacturing. ACS Biomater Sci Eng **4**, 2292–2307 (2018)
11. P.J. Bártolo et al., Biomanufacturing for tissue engineering: Present and future trends. *Virt Phys Prototyp* **4**(4), 203 (2009)
12. P.S.P. Poh et al., Polylactides in additive biomanufacturing. Adv Drug Deliv Rev **107**, 228–246 (2016)

13. Smart biomanufacturing: next-generation biopharmaceutical production, https://new.siemens.com/us/en/markets/pharma-industry/smart-bio.htm
14. C. Slouka et al. (eds.), *Continuous biomanufacturing in microbial systems* (Frontiers Media, Lausanne, 2021)
15. Biomanufacturing: How biology is driving manufacturing into the future, https://www.advancedtech.com/blog/future-of-biology-in-manufacturing/
16. R. A. Rader, Top trends in biomanufacturing, February 2021, http://www.processdevelopmentforum.com/articles/top-trends-in-biopharmaceutical-manufacturing-2017/
17. Embracing innovation in biomanufacturing, https://bioprocessintl.com/manufacturing/single-use/embracing-innovation-in-bioprocessing-and-biomanufacturing/
18. New decade, old challenge: Biomanufacturing workforce development remains key to industry growth, https://biobuzz.io/new-decade-old-challenge-biomanufacturing-workforce-development-remains-key-to-industry-growth/
19. S. Pearson, Can the U.K. maintain its biomanufacturing empire? December 27, 2013., https://www.genengnews.com/insights/can-the-u-k-maintain-its-biomanufacturing-empire/
20. Manufacturing and biomanufacturing: Materials advances and critical processes, https://www.nist.gov/system/files/documents/2017/05/09/manufacturing_biomanufacturing_wp_08_11.pdf
21. Biomanufacturing: A path to sustainable economic recovery, July 2020, https://www.newstatesman.com/spotlight/emerging-technologies/2020/07/biomanufacturing-path-sustainable-economic-recovery
22. How microbial fermentation and protein biomanufacturing works, https://luinabio.com.au/how-microbial-fermentation-and-protein-biomanufacturing-works/
23. Biomanufacturing in Canada, https://www.ic.gc.ca/eic/site/151.nsf/eng/home
24. Global evolution of biomanufacturing, https://bioprocessintl.com/manufacturing/monoclonal-antibodies/global-evolution-of-biomanufacturing-340616/
25. S. Henrikssen, Samsung biologics to invest $2B in 'super plant' for biomanufacturing, November 2020, https://realitysandwich.com/samsung-biologics/
26. Biomanufacturing, https://www.a-star.edu.sg/Research/research-focus/biomanufacturing
27. Six key trends influencing the next-generation biomanufacturing market, August 2020, https://blog.bisresearch.com/six-key-trends-influencing-the-next-generation-biomanufacturing-market
28. Y.P. Zhang, J. Sun, Y. Ma, Biomanufacturing: history and perspective. J Ind Microbiol Biotechnol **44**, 773–784 (2017)
29. D. G. Jensen, The biomanufacturing career track, July 2009, https://www.sciencemag.org/careers/2009/07/biomanufacturing-career-track
30. C. Prakash et al., *Biomanufacturing* (Springer, 2019)
31. G. Subramanian (ed.), *Continuous biomanufacturing: innovative technologies and methods* (John Wiley & Sons, 2017)
32. P.L. Show, K.W. Chew, in *The prospect of Industry 5.0 in biomanufacturing*, ed. by T.C. Ling, (Taylor & Francis, 2021)
33. N.J. Smart, *Lean biomanufacturing: creating value through innovative bioprocessing approaches* (Woodhead Publishing, 2013)
34. J. Odum, M.C. Flickinger (eds.), *Process architecture in biomanufacturing facility design* (John Wiley & Sons, 2017)
35. J.J. Zhong (ed.), *Biomanufacturing* (Springer, 2004)
36. E.Y. Lee (ed.), *Recent advances in biocatalysis and metabolic engineering for biomanufacturing* (MDPI, Basel, 2019)
37. P.L. Show, K.W. Chew, T.C. Ling (eds.), *The prospect of industry 5.0 in biomanufacturing* (CRC Press, Boca Raton, FL, 2021)
38. P. Calvert, R. Narayan (eds.), *Computer aided biomanufacturing* (John Wiley & Sons, 2011)

Chapter 19
Future of Manufacturing

The future of manufacturing is in technology. The next generation of manufacturing champions will come from those companies that use brain power, not labor, to drive their innovations.

–Steven Moore

19.1　Introduction

In essence, manufacturing is the act of transforming raw material into products that benefit society. It is a key component of economic growth and the bedrock on which a nation's economy is built. However, manufacturing is constantly changing and continuously evolving from concept development to methods and tools. It is no longer just a process that turns raw materials into physical products. It consists of a wider set of activities that create value and benefits for the society. It contributes to wealth, exports, innovation, and productivity growth. As manufacturers desire to stay competitive in the marketplace, they are constantly searching for the latest and greatest inventions, strategies, and systems. They welcome technology that promises greater efficiency and productivity with open arms [1].

The major challenges limiting the potential sales and profit growth in the manufacturing section are market volatility, rising material costs, product quality and reliability, price reduction pressures, increasing labor costs, digitizing the workforce, level of digitization, creativity and innovation, and transportation/logistics costs. Also, the current manufacturing infrastructure is costly to maintain, integrates poorly with other systems, and prevents companies from capitalizing on the advantages of digitalization. To address these challenges and make a successful transition to the next generation of manufacturing, manufacturers are making significant investments in equipment and systems and adoption of new technologies [2]. Manufacturers always strive to be more innovative, whether by developing new products or improving existing ones. Manufacturing is increasingly becoming more efficient, intelligent, customized, sustainable, and automated. The future of manufacturing sector is undeniably digital and will be shaped by a combination of advancements in technology.

Technology is evolving and is transforming the manufacturing industry. Each year brings new technologies that are changing the face of manufacturing. Some of the most significant benefits that technology brings to the manufacturing industry include energy efficiency, predictive maintenance, improved product quality, downtime reduction to zero, increased level of digitization, enhanced customer demand, and informed decisions [3]. Emerging technologies that have great impact in manufacturing include artificial intelligence, smart manufacturing, robotic automation, 3D printing, nanotechnology, industrial Internet of Things, and augmented reality. The use of these technologies will have a profound impact on the future of manufacturing industry. They have the potential to transform manufacturing as we know it. They should be at the core of any manufacturing upgrading effort [4].

This chapter takes a close look at the challenges that manufacturers are facing now and see what the future holds for manufacturing. It begins by reviewing the emerging manufacturing technologies. It considers the future of technology in manufacturing and the factory of the future. It highlights the challenges facing the manufacturing industry in the future. It covers the future of global manufacturing. The last section concludes with comments.

19.2   Emerging Manufacturing Technologies

All manufacturers can position themselves to capitalize on the benefits of new and innovative emerging technologies. By prioritizing a digital-first strategy and planning holistically around key issues, leaders can regear their companies' all-round capabilities and competitiveness. They must also wrestle with disruptive technologies that have been appearing on the horizon for years now. These include [5]:

- Social and mobile capabilities were some of the first next-generation trends that became must-have capabilities very quickly.
- Product configuration tools such as the NX and CATIA turnkey integrated design, manufacturing and enterprise-wide systems, incorporating integrated computer-assisted design (CAD) systems, and late-stage assembly helped manufactures meet demands for mass customization.
- Robotics changed operational technology dramatically, just as shop-floor automation revolutionized workflows and production cycles.
- Big data started conversations about factories becoming predictive, leveraging analytics, and creating strategies for identifying customer expectations.
- Cloud computing fueled the imagination further, making storage of immense pools of data possible.
- Prototyping with 3D printing (or additive manufacturing) tooling allows manufacturers to design and print their own tools.
- The Internet of Things far surpasses all of these disruptive technologies. The potential impact of the IoT is truly staggering and hard to grasp.
- Nanomaterials will be used more widely in the future of manufacturing.

- Manufacturing workers will use AI to speed the adoption of technologies augmenting their work.
- Manufacturing will transition from making parts to "as-a-service" business models.
- New materials pose new requirements on the manufacturing plants.

This is not a comprehensive list. These technologies can connect billions of people to the web and drastically improve the efficiency of business and organizations. Some of the technologies will be further discussed in the next section.

19.3 Future of Technology in Manufacturing

The future of manufacturing is essentially in everybody's interest. It will determine our jobs, our consumption patterns, and the quality of life for this and future generations [6].

Nanotechnology, cloud computing, robotics, artificial intelligence, the Internet of Things, Industry 4.0, and additive manufacturing are all improving manufacturing processes. It is worth examining the common thread running through all these technologies [7, 8].

- *Robotic Automation:* Automation is vital aspect of the manufacturing industry's future. Some consider it as the future of manufacturing. "Automation for all" is regarded as the next step in the industry. Manufacturers of all sizes are embracing various forms of automation in order to lower costs, increase production, and reduce response times. Robots are often used in high-production, high-volume environments. Robotic automation has found its way into many areas of manufacturing such as welding, assembly, shipping, handling raw materials, product packing, and shipping. It offers manufacturers opportunities to save on costs, enhance production, and remain competitive. It also offers manufacturers growing opportunities to save on costs, enhance production, and remain competitive. In the manufacturing arena, robots have become smarter, safer, and more mobile. Robots have been designed with incredible skills like increasing dexterity or machine vision. They will increase efficiency and provide greater benefits to their human coworkers. Manufacturing can rely on a mix of human and robotic labor which will increase productivity and ROI. The most well-known example of robotic automation is the Fanuc plant in Japan, where robots build other robots. Fig. 19.1 shows typical robotic automation [9].
- *Industry 4.0:* This is shorthand for the fourth industrial revolution, the previous three being mechanical production, mass production, and the digital revolution. Industry 4.0 encompasses new technologies that combine the physical, digital, and biological worlds, impacting all disciplines, economies, and industries. The term is directly related to manufacturing and could be called Manufacturing 4.0. It has revolutionized the manufacturing sector by providing manufacturers with opportunities to utilize advanced tools and technologies throughout the product

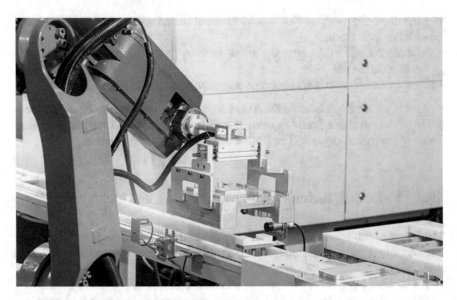

Fig. 19.1 A typical robotic automation [9]

lifecycle. It remains a popular topic when discussing the current and future state of manufacturing. It allows more flexible production, greater productivity, and the development of new business models. The components of Industry 4.0 are displayed in Fig. 19.2 [10]. Industry 4.0 is driving the manufacturing industry to modernize factories. Whether you call it Industry 4.0, the industrial Internet of Things (IIoT), or smart manufacturing, manufacturing is in the midst of a significant transformation.

- *Internet of Things* (IoT): This is the answer for manufacturers looking for solutions to complete maintenance proactively. IoT refers to the connectivity between devices that are able to communicate and exchange data with one another. IoT is revolutionizing manufacturing. It may be regarded as the nerve system of a manufacturing plant. It is helping manufacturers connect and monitor the various components of their operations, gaining insight never before possible. The IoT market is growing steadily and is likely to continue to do so in the foreseeable future. The industrial Internet of Things (IIoT) is the future of manufacturing. It refers to the use of a system of interconnected computing devices that can transfer data over a network without any human involvement. The idea is that everything, from the bottom up to the cloud, needs to be connected.

- *Additive Manufacturing or 3D Printing:* This technology is rapidly gaining traction and revolutionizing how the manufacturing industry approaches product development. Manufacturers will benefit from faster, less expensive production as a result of 3D printing. Advancements in additive manufacturing can help companies reduce energy costs, limit waste, and boost production. 3D printing gives manufacturers a way of creating tangible product prototypes without committing to a costly production process.

Fig. 19.2 Components of Industry 4.0 [10]

- *Artificial Intelligence:* AI can convert large amounts of raw data into information that a human can read and interpret. AI has played an increasingly vital role in automation. AI-powered robots are increasingly being used in manufacturing. Manufacturers are attempting to have their manufacturing systems partially or fully automated in order to reduce labor costs, increase operation efficiency, and increase the production rate. The integration of the expert system, an artificial intelligence tool, into mainstream information technology will promote increased exploitation of this technology in the manufacturing sector.
- *Virtual Reality and Augmented Reality:* Virtual reality is becoming very popular within the entertainment industry as it is allowing consumers to immerse themselves in the worlds in which they play. Augmented reality gives manufacturers the opportunity to test products in a virtual real-life platform. Augmented reality is also a tool that manufacturers will find useful in increasing productivity. Augmented reality (AR) and virtual reality (VR) have different applications in manufacturing and are poised to improve manufacturing in myriad ways. These

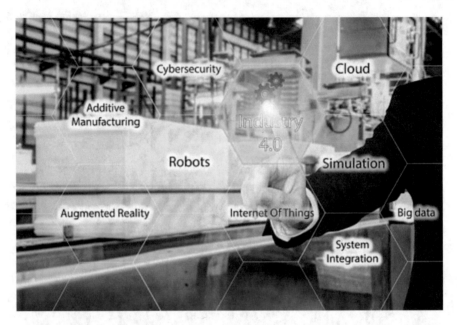

Fig. 19.3 Emerging technologies in manufacturing [11]

technologies will impact areas like training and maintenance in various ways. They will yield greater efficiency, improved accuracy, and predictive maintenance by providing more visibility and insight into processes.

These emerging technologies are illustrated in Fig. 19.3 [11]. They are quickly becoming the future of manufacturing. The convergence of these is incredibly important to the future of manufacturing. Manufacturers must keep eye on these disruptive technologies, which will slowly make their way into more and more facilities and collectively determine the future of manufacturing.

Other important technologies such as nanotechnology, cloud computing, business intelligence, and blockchain are all improving manufacturing. They are pushing manufacturers to explore radically new ways of creating and capturing value. For both incumbents and new entrants, these technologies will serve as tools to successfully navigate the new landscape of manufacturing. Nations that are currently investing in these technologies are the ones that will dominate global manufacturing in the future. Although manufacturing will be dominated by new technologies, the changes will be led by humans.

19.4   Factory of the Future

Many manufacturers are beginning to modernize their factory operations to achieve greater efficiency, flexibility, and productivity. This modernization process requires a substantial upgrade of the existing software infrastructure. Figure 19.4 shows a

Fig. 19.4 A typical factory of the future at Hannover Messe, USA [12]

typical factory of the future [12]. The factory of the future has evolving definitions or names such as smart manufacturing, Industry 4.0, or the digital enterprise.

For sure, the factory of the future will be interconnected, flexible, smart, small, and movable (to where the resources are and where the customers are) with workers, customers, and suppliers all having more real-time transparency. Future factories will make intensive use of automation, technology, and people [13]:

- *Automation:* Automation makes it possible for manufacturers to reach new levels of production, precision, and safety. It will become even more influential in the manufacturing industry. It is predicted to drive job growth in the manufacturing industry. In the factory of the future, robots will be far more advanced and collaborative. Robots have become invaluable for monotonous jobs such as packaging, sorting, and repeated lifting. The widespread misperception that automation and AI will put manufacturing jobs at risk is not valid.
- *Technology:* The factory of the future will employ a wide range of cutting-edge, fast-changing, disruptive technologies. Factories will make use of modern technologies such as artificial intelligence, virtual and augmented reality, and the Internet of Things. These technologies will be integrated within the factory.
- *People*: The future of manufacturing is people. It's technology-enabled but human-driven. In the future factory, there is a need for a highly talented, skilled, and flexible workforce. Workers in a highly-skilled manufacturing sector are a true competitive asset for any company. Manufacturing decision-makers must create a friendly environment in order to get the most out of employees.

19.5   Challenges

Technological advances are not without their challenges. Some of the challenges that the manufacturing industry will face in the future include [14]:

- *Global Competition:* The growing trade war between China and the United States continues to strain manufacturing worldwide. According to the 2019 US-China Business Council survey on trade relations between the two countries, the top concern among foreign companies operating in China was unfair competition. China has become the world's leading manufacturing country. *Manufacturing in the United States has been in long-term decline, with the competitive advantage moving to low-cost countries such as Mexico and China. This has caused the United States to change its focus and* move to innovative products that require engineering and sophistication. With increasingly automated factory, the advantage of being a low-cost country (such as Mexico, China, and Thailand) starts to fade.
- *New Technology:* Technology is advancing at a rapid pace. Technology is changing the comparative advantages that drive competitiveness. It is no secret that technology is rapidly growing, and companies across all industries are understanding how to utilize the best tools for bottom-line success. Due to the complex nature of manufacturing, the adoption of latest technologies is always difficult. A tough decision is allocating resources toward R&D or sustaining their existing systems.
- *Customer Demand:* Customers demand personalization and customization from manufacturers. Increasing consumer demand for high-quality, personalized goods is fueling the need for flexible and fast manufacturing processes that respond to customer and market demands. The shifts in consumer demand are compelling manufacturers to explore radically new ways of creating and capturing value. Manufacturers will need to deal with the increased customer demand for personalization, faster completion, and maximum transparency.
- *Cybersecurity:* This should be a major concern for all industries including manufacturing. Connected devices are vulnerable to cyberattacks. Cybersecurity is a national and global phenomenon of grave importance because the malicious use of cyberspace could hamper economic, public health, safety, and national security activities. It is becoming more and more important as more information is being made available on computer networks. Cybersecurity breaches cannot be stopped at a nation's borders since it is difficult to determine where the actual borders are in cyberspace. Thus, cybersecurity can become a supranational problem. Manufacturers will need to mitigate risks more than ever [15].
- *Workforce:* An important question to ask is what will the manufacturing workforce in the future look like? The future of manufacturing is people. The manufacturing sector will still need people who can manage new operations, manage robotics, and maintain them. The future of manufacturing depends on building

the necessary workforce. Manufacturers are reporting talent shortages as they struggle to find the right blend of technical and soft skills to fill new positions. The mismatch between available workers and the advanced technical skills necessary in modern manufacturing has been dubbed the "skills gap" [16]. The gap between available jobs and the number of skilled workers is growing. This skills gap remains a serious problem. As the manufacturing landscape changes with time, workers' roles in the manufacturing industry will change. Manufacturers will need more people with technological expertise. In other words, manufacturers still need a large number of skilled workers to handle new technologies. Instead of detracting from the role humans play in manufacturing, AI will support them to take on new roles, ultimately creating more opportunities for employees than it will eliminate [17]. The manufacturing workforce will shift in size, shape, preparation, and required skills.

Figure 19.5 illustrates these challenges. Meanwhile, the boundary separating product makers from product sellers is increasingly permeable.

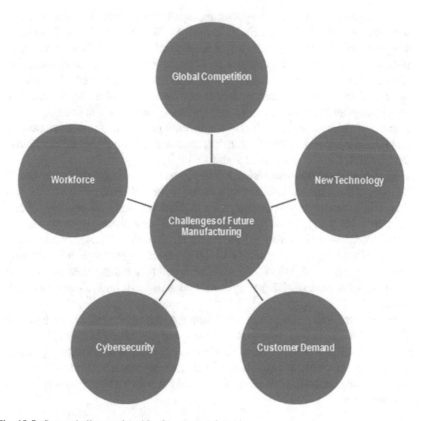

Fig. 19.5 Some challenges faced by future manufacturing

19.6   Future of Global Manufacturing

Technology is changing so fast that what was successful today is not necessarily going to be as successful tomorrow. Globally, manufacturing continues to grow and thrive. The success of a manufacturing business depends on how quickly it can adapt to the changing world. The more agile the business operations can be, the faster they can adapt to this new world. Although the global manufacturing sector has undergone a tumultuous decade, manufacturing is critically important to both developing and developed nations. It improves standards of living, builds economic prosperity, provides employment, promotes innovation, increases productivity, minimizes human errors, enhances efficiency, and boosts return on investment (ROI). Globally, manufacturing continues to grow and change. The United States, Europe, and East Asia dominate global manufacturing through heavy investment in robotics, AI, and IoT technologies. Economies such as India, Indonesia, and China have risen to the top ranks of the 15 largest global manufacturing economies [18]. We now consider the future of manufacturing in some nations.

- *United States:* America's manufacturing revolution is occurring now. Although manufacturing in the United States as well as "Made in the USA" has been in long-term decline, US manufacturers are typically bullish about future growth prospects. Emerging technologies such as nanotechnology, automation, and robotics have modernized American factories making them smaller, modular, safer, and smarter than their predecessors. Manufacturers in the United States should have an opportunity to win more business there while continuing to compete on the global stage for highly engineered products. Although some claim that China is becoming more expensive, the competitive advantage may not necessarily shift back to the United States.
- *United Kingdom:* Manufacturing continues to be an essential part of the UK economy. The importance of manufacturing to the UK economy is incontrovertible. UK government demonstrated its support of manufacturing by uniquely commissioning a strategic look at the future of manufacturing as far ahead as 2050. Constant adaptability will pervade all aspects of manufacturing in the United Kingdom, from research and development to customer interdependencies. The quality and skills of the workforce will be a critical factor in capturing competitive advantage. UK policymakers need to focus on the supply of skilled workers [19].
- *India:* Prime Minister Narendra Modi launched a program called Make in India. Do you think initiatives like Make in India and the ascent of manufacturing in other parts of Asia have a chance of succeeding? A large paint company in India has used the combination of industrialized sensors, automation, and social media analytics to catalyze both internal operations and customer-driven product development.

- *Asia:* Asia has been known as a cost-effective manufacturing hub due to its vast labor pool. Several Asian manufacturers fabricate products ranging from automobiles to consumer electronics. As the advantage of labor dwindles, newer paradigms of productivity, operational visibility, and risk management are needed for Asian manufacturers to stay competitive. By getting rid of their traditional approaches to maintaining competitiveness, Asian manufacturers can fully capture the benefits of digital. Manufacturers in Asia need to upskill their workers, focusing on developing and capitalizing on the capabilities that are uniquely human. They need to act today or risk becoming obsolete [20].
- *Latin America:* China and its "unlimited supply of labor," rapid productivity, growth, scale, and government intervention has made China a formidable competitor to Latin American manufacturers and raises many questions about their future. This has unprecedentedly challenged the future of Latin America and the Caribbean (LAC) in the world's division of labor. Mexico and Brazil are apparently on the way to a head-on collision with Chinese exporters since they rely heavily on industries such as apparel, textile, and electronics where China has huge comparative and competitive advantages. The future of manufacturing in LAC is often seen with pessimism on the grounds of geography and endowments. Chinese challenge does not make the manufacturing future any brighter in LAC [21].

19.7 Conclusion

Although no one can predict the future, one can confidently say that the future of manufacturing is bright. The future of manufacturing is not only technology-enabled but human-driven. The companies that can adapt, innovate, and utilize global resources will generate significant growth and success. Large-scale production will continue to dominate some segments of the value chain. The future of manufacturing looks bright.

Manufacturers are always confronted with a choice: to adopt new technology and move forward or ignore it and stand still. Forward-looking manufacturers know that, to stay on top, they must keep ahead of industry trends. They are already revisiting and shaping their manufacturing capabilities.

Since the future of manufacturing is human, there is a need to prepare the next generation of skilled workers in a future manufacturing environment that incorporates increased use of artificial intelligence, robots, 3D printing, and biotechnology. Future workers must have a combination of both hard (technical) skills and soft skills to succeed. Both governments and businesses should make education and the development of skills their priorities. Manufacturers must constantly train their workers to keep them current and become citizen developers. For more information about the future of manufacturing, one should consult the books in [22–24].

References

1. M.N.O. Sadiku, O.D. Olaleye, A. Ajayi-Majebi, S.M. Musa, Future of manufacturing. Int J Trend Res Develop **8**(1), 194–197 (2021)
2. The future of manufacturing: 2020 and beyond, https://www.kronos.com/resources/future-manufacturing-2020-and-beyond
3. What does the future of manufacturing technologies look like in 2019? June 2019, https://www.manufacturing.net/industry40/article/13251981/what-does-the-future-of-manufacturing-technologies-look-like-in-2019
4. M.N.O. Sadiku, T.J. Ashaolu, A. Ajayi-Majebi, S.M. Musa, Emerging technologies in manufacturing. *International Journal of Scientific Advances* **1**(2) (2020)
5. L. Korak, Preparing for the factory of the future, April 2016, https://www.industryweek.com/technology-and-iiot/emerging-technologies/article/21972483/preparing-for-the-factory-of-the-future
6. A. Primer, The future of manufacturing and development: three things to remember, 2018., https://oecd-development-matters.org/2018/06/18/the-future-of-manufacturing-and-development-three-things-to-remember/
7. Top 7 manufacturing trends for 2020, https://www.advancedtech.com/blog/top-7-manufacturing-trends-for-2020/
8. The future of manufacturing, https://www.airedalesprings.co.uk/the-future-of-manufacturing/
9. Automation: The future of manufacturing is here, https://www.qualitymag.com/articles/96158-automation-the-future-of-manufacturing-is-here
10. The future is…Industry 4.0, https://www.materialise.com/en/blog/future-industry-40
11. J. P. Donlon, Building a better road map to the future of manufacturing, May 2017, https://chiefexecutive.net/building-better-road-map-future-manufacturing/
12. Tetra Pak introduces the 'factory of the future' with human and AI collaboration at its core, April 2019, https://plas.tv/?p=12467
13. S. Griffiths, The future of manufacturing technology, February 2914, https://www.syspro.com/blog/erp-for-manufacturing/the-future-of-manufacturing-technology/
14. D. Evans, Looking ahead: the future of manufacturing in 2020 and beyond, 2019., https://www.fictiv.com/articles/manufacturing-in-2020-and-beyond
15. M.N.O. Sadiku, S. Alam, S.M. Musa, A primer on cybersecurity. Int J Adv Scientif Res Eng **3**(8), 71–74 (2017)
16. D. Bolin, 2020 vision: A look ahead at the future of manufacturing, https://amatrol.com/2020-vision-future-of-manufacturing/
17. Where do people fit in the factory of the future? https://zenoot.com/articles/where-do-people-fit-in-the-factory-of-the-future/
18. McKinsey Global Institute, Manufacturing the future: The next era of global growth and innovation, November 2012, https://www.mckinsey.com/~/media/McKinsey/Business%20Functions/Operations/Our%20Insights/The%20future%20of%20manufacturing/MGI_Manufacturing%20the%20future_Executive%20summary_Nov%202012.pdf
19. The future of manufacturing: A new era of opportunity and challenge for the UK, https://assets.publishing.service.gov.uk/government/uploads/system/uploads/attachment_data/file/255923/13-810-future-manufacturing-summary-report.pdf
20. Industrial Automation Asia, The digital future of manufacturing: Are Asian manufacturers ready? https://www.iaasiaonline.com/digital-future-manufacturing-asian-manufacturers-ready/
21. M.M. Moreira, Fear of China: is there a future for manufacturing in latin America? World Dev **35**(3), 355–376 (2007)
22. R. McCormack et al. (eds.), *Manufacturing a better future for America* (Alliance for American Manufacturing, 2009)
23. M. Hallward-Driemeier, G. Nayyar, *Trouble in the making?: the future of manufacturing-led development* (World Bank Group, Washington DC, 2018)
24. National Research Council, *Visionary manufacturing challenges for 2020* (National Academy of Sciences, 1998)

Index

A

Additive biomanufacturing, 264
Additive manufacturing, 3, 29, 71,
        109–120
    applications of, 113
    benefits of, 115
    challenges of, 116
    global adoption of, 118
    processes of, 112
Additive nanomanufacturing, 206, 223, 278
Advanced manufacturing, 28, 219
Aerospace, 111
Affordability, 267
Agility, 59
AI tools, 17
Artificial intelligence (AI), 2, 13–31
    benefits of, 24
    branches of, 18
    challenges of, 25
    in manufacturing, 17, 19
    overview of, 14
Asia, 285
Asset management, 90
Augmented reality, 6, 71, 279
Australia, 211, 269
Automated production lines, 40
Automation, 20, 36, 43, 88, 114, 191, 222,
        271, 281
    overview on, 35
Automotive industry, 20, 39, 113, 133,
        147, 164
Autonomy, 59

B

Big data, 71, 95–106
    applications of, 100
    benefits of, 102
    challenges of, 103
    characteristics of, 96
    in manufacturing, 99–100
    standardization of, 104
Big data analysis (BDA), 98–99, 250
Biomanufacturing, 259–273
    applications of, 262
    benefits of, 267
    challenges of, 267
    future of, 270
    global adoption of, 268
Biopharmaceutical industry, 262

C

Canada, 45, 152, 269
Cellular manufacturing, 163
Cement industry, 132
Chemical industry, 56
Chemical manufacturing, 148
China, 26, 45, 63, 79, 92, 137, 153, 181, 196,
        240, 269
Cloud, 186, 188
    migrating into, 190
Cloud computing, 71, 186, 188
    characteristics of, 217
    deployment models of, 187
    overview of, 186

Cloud manufacturing, 185–198, 219
   applications of, 191
   architecture for, 189
   benefits of, 192
   challenges of, 194
   future of, 197
   global adoption of, 194
Cloud robotics, 191
Collaboration, 271
Common fear, 25
Competitiveness, 167
Complexity, 25, 267
Computer-aided-design (CAD), 111
Computer industry, 132
Connectivity, 58
Construction industry, 147
Contract manufacturing, 236
Cost, 43, 117, 209, 224, 237
Cost-effectiveness, 192
COVID-19 pandemic, 9
Customer experience, 101
Customer satisfaction, 89, 167
Customization, 177, 225
Customized manufacturing, 148
Cyber manufacturing, 247–257
   attacks on, 251
   benefits of, 254
   concept of, 248
   enabling technologies of, 249
   global adoption of, 255
Cyber-physical systems, 51, 55, 72, 249
Cybersecurity, 71, 104, 247, 248, 282
   handling risks of, 253

D
Data, 95, 99
   in manufacturing, 99, 100
Data breaches, 194
Data scientists, 98
Data security, 244
Decentralization, 60, 74
Deep learning, 16
Denial-of-service attacks, 252
Distributed manufacturing, 175–183
   applications of, 178
   benefits of, 180
   challenges of, 180
   concept of, 176
   features of, 177
   global adoption of, 180
Distributed manufacturing systems, 176
Drug delivery, 207

E
Efficiency, 43, 74, 225
Electronics, 206, 265
Electronics manufacturing, 39, 114, 191
Emerging technologies, 2, 276
   benefits of, 8
   challenges of, 9
   future of, 277
   global aspect of, 9
Energy, 114, 133
Energy system, 57
Enterprise resource planning
   (ERP), 187
Environmental friendliness, 134
Environmental health, 149
Environmental impact, 25
European Union, 137, 201, 229
Expert system, 14
Explainable AI, 29

F
Factory of the future, 280
Feynman, Richard, 202
Flexibility, 74, 115
Food industry, 114
Food manufacturing, 264
Ford, Henry, 67
Furniture manufacturing, 147, 178
Future of manufacturing, 275
   challenges of, 282
   global type of, 281
Fuzzy logic, 14

G
Germany, 6, 27, 63, 79, 105, 118, 256
Ghana, 138
Global competition, 157
Globalization, 244
Green, 126
Green design, 129
Green disposal, 130
Green manufacturing, 123–138, 163
   applications of, 131
   awareness of, 133
   benefits of, 133
   characteristics of, 129
   global impact of, 136
   motivations for, 127
   practices of, 129
Green production, 129
Green resources, 129

H
Healthcare, 205
Hospital, 166
Human factors, 194

I
Identity theft, 252
IEEE Future Directions Committee, 9
India, 27, 63, 92, 138, 153, 171, 241, 284
Industrial Internet, 86
Industrial Internet of Things (IIoT), 4,
     55, 85, 87
Industrial revolutions, 68, 201
Industry 4.0, 6, 21, 67–80, 162, 163, 219, 277
     applications of, 74
     benefits of, 76
     challenges of, 77
     features of, 72
     global adoption of, 79
     overview on, 69
     principles of, 72
Integrated manufacturing technology, 220
Intellectual property (IP), 238, 244
Intelligent manufacturing, 23, 75
Intelligent robots, 70
Internet of Things (IoT), 55, 71, 83–93,
     249, 278
     applications, 87
     benefits of, 89
     challenges of, 90
     overview on, 84
Interoperability, 60, 73, 91
Iron and steel industry, 132

J
Japan, 45, 137, 169, 229
Job loss, 243

K
Korea, 229

L
Lack of skills, 91
Lack of standard, 227
Latin America, 285
Lean management, 165
Lean manufacturing, 157–171
     applications of, 164
     benefits of, 166

concept of, 159
global adoption of, 169
principles of, 160
Liability, 25
Lights-out manufacturing, 40
Limited commercialization, 209
Localization, 177, 180
Local manufacturers, 237

M
Machine learning, 2, 15
Machine tool industry, 165
Machining industry, 191
Malware, 251
Man-in-the-middle attacks, 252
Manufacturers, 1, 95
Manufacturing, 1, 10, 13, 18, 33, 49, 84, 95,
     123, 141, 183, 201, 215, 218, 230, 233,
     247, 259, 275
     future of AI in, 26
     global AI in, 26
Manufacturing executive systems (MES),
     224, 251
Manufacturing industry, 18, 67, 83, 95, 136,
     150, 185, 205, 247
Manufacturing plants, 88
Medicine, 114, 262
Mexico, 241
Military, 114
Miniaturization, 207
Mining industry, 76, 88
Modern computing, 216
Modern manufacturing, 64, 185
Modularity, 74

N
Nanomachines, 206
Nanomanufacturing, 201–212
     applications of, 205
     benefits of, 207
     challenges of, 208
     future of, 211
     global adoption of, 209
     methods of, 205
     nanomaterials in, 205
Nanomaterials, 205, 206
Nanotechnology, 8, 201, 203, 207
     overview of, 202
Nanoworld, 201
NASA, 3
Natural language processing, 16

Neural networks, 15
Norway, 153

O
Offshore locations, 240
Offshoring, 233, 234239
Offshoring manufacturing, 233–244
  benefits of, 242
  challenges of, 242
On-shore manufacturing, 178
  definition of, 234
Optimization, 59
Outsourcing, 239, 240
Overexposure, 209

P
Personalization, 177
Pharmaceutical manufacturing, 148
Pharmaceuticals industry, 132
Phishing, 252
Plastic industry, 165
Pollution reduction, 133, 134
Predictive maintenance, 23, 75, 100
Predictive manufacturing, 22
Printing industry, 165
Productivity, 43, 89
Prototyping, 115, 193

Q
Quality, 165, 167

R
Radio frequency identification (RFID), 55, 221
Ransomware, 252
Remanufacturing, 128, 148
Robot, 17, 34, 35, 55, 221
  uses of, 38
Robotic automation, 2, 33, 277
  benefits of, 43
  challenges of, 44
  global adoption of, 45
  in manufacturing, 33
Robotic processing automation, 41
Robotics, 17, 19, 24, 34
  in manufacturing, 37
Russia, 270

S
Scalability, 192
Security, 91, 117, 191, 226
Semiconductor manufacturing, 22, 56, 133
Sensors, 55, 221
Sewing industry, 165
Shortage in workforce, 25
Shortage of skills, 237
Simulation, 72
Singapore, 182, 270
Six Sigma, 162
Smart automotive manufacturing, 88
Smart biomanufacturing, 265
Smart factory, 21, 58, 75, 223, 225
Smart machining, 57
Smart manufacturing, 5, 6, 21, 28, 49–65, 219
  applications of, 56
  benefits of, 60
  challenges of, 61
  characteristics of, 52
  concept of, 51
  enabling technologies of, 54
  global adoption of, 62
  goals of, 54
Social engineering attacks, 251
Societal impact, 135
Southeast Asia, 119
South Korea, 270
Spain, 182
Standardization, 104, 169
Standards, 91
Supply chain management, 23, 88, 101
Sustainability, 192, 267
  concept of, 142
  types of, 143, 144
Sustainable approaches, 146
Sustainable construction, 178
Sustainable Development Goals, 151
Sustainable logistics, 148
Sustainable manufacturing,
  141–153
  benefits of, 149
  challenges of, 150
  examples of, 147
  global adoption of, 151

T
Technical limitations, 117
Technology, 276, 281, 282

Textile industry, 132
3P printing, 3, 4, 55, 278
Tissue engineering, 263, 264
Traditional manufacturing, 13, 49, 67, 124
    overview of, 50
Transparency, 59
Transportation, 132

U
Ubiquitous computing, 216, 217, 222
Ubiquitous connectivity, 224, 257
Ubiquitous manufacturing, 215–231
    applications of, 222
    benefits of, 224
    challenges of, 226
    concept of, 218
    enabling technologies of, 221
    future of, 229
    global adoption of, 228
Ubiquitous technology, 220

United Kingdom, 63, 92, 105, 118, 152, 171,
    196, 211, 256, 269, 284
United States, 26, 45, 63, 79, 92, 105, 118,
    137, 151, 170, 181, 195, 209, 228, 255,
    268, 284

V
Value, 160
Virtualization, 60
Virtual reality, 279

W
Waste, 158
Waste minimization, 125
Weiser, Mark, 216

Z
Zimbabwe, 171